Studies in Smart Technologies

Advanced communication and computing technologies to develop intelligent and adaptive systems are usually referred to as smart technologies. Many of the sustainable development goals (e.g., SDG 4, SDG 6, SDG 7, SDG 8, SDG 9, SDG 11) cannot be achieved without developing smart technologies. The aim of the proposed book series, "Studies in Smart Technologies," is to provide a detailed and latest overview of the latest research findings in the field of smart technologies and allied areas. The series covers studies related to smart cities, IoT, smart sensors, AI, soft computing, and other related areas. The content published in the series will be an important reference to the policymakers, industries, researchers, practitioners, and students in computer science, electrical engineering, information technology, and aerospace engineering.

The series will publish monographs, edited volumes, textbooks and proceedings of important conferences, symposia, and meetings in the field of smart technologies and their applications.

Eleni G. Makri

Metaverse and Smart Grid in Energy Industry: An Open Innovation Challenge on Trends and Applications

Springer

Eleni G. Makri
Faculty School of Humanities-Business
UNICAF
Larnaca, Cyprus

Warnborough College
Canterbury, Kent, UK

ISSN 3004-9881 ISSN 3004-989X (electronic)
Studies in Smart Technologies
ISBN 978-3-032-29665-8 ISBN 978-3-032-29666-5 (eBook)
https://doi.org/10.1007/978-3-032-29666-5

This Springer imprint is published by the registered company Springer Nature Switzerland AG
The registered company address is: Gewerbestrasse 11, 6330 Cham, Switzerland

In loving memory of my parents, doctor Georgios Makris and teacher Georgia Tsiotou-Makri.

Preface

Governments, public agencies, non-profit organizations, academia, industry, communities of practice and citizens actively share their assets, resources and human capital to support a United Nations (UN)-driven sustainable society, effectively. Escalating climate crisis events are disruptive across the world with emergent life-threatening, environmental, economic, social and urban effects. The envisioned Metaverse ecosystem and emerging smart grid (i.e., intelligent) technologies are powerful, revolutionary drivers that share the agency and capacity to address the worldwide, demanding challenges of sustainable energy generation, maintenance, transmission and consumption, efficiently. This envisaged Metaverse-driven, open innovation-led distributed global intelligent energy network (i.e., inflow and outflow innovation) encompasses diverse trends, applications and interventions built upon renewable resources. This forecasted, eco-friendly ecosystem blends the system's design thinking (i.e., open innovation) with extended and deep-level, sustainable, smart energy production and deployment interventions that can be exercised on Metaverse. The core premise of this book is to highlight the need or insight to expand our current limited awareness, insights, reflection, research data and actionable practice on how best to use the intelligent (smart) innovations to support an envisioned, eco-and-open innovation Metaverse energy industrial sector, "echoing" the connected multimodal challenges (i.e. eco-sustainability, scalability, affordability, safety, monitoring, regulatory standards, etc.). This predicted open-and-eco "SmartEnergyverse" network is intertwined across the following pillars, as next.

Chapter 1 sketches our envisioned outline of present, yet rare but progressing open research and enterprise challenging issues that align with the "next frontiers", open innovation-led operationalization of the future industrial energy Metaverse and smart grid(s). This framework is built upon dispersed and "greener" intelligent technologies and energy agents, assets, resources and operations (i.e. fog-and-edge, sensing, data analytics, production, maintenance, transmission, consumption, etc.). Chapter 2 enlightens additional modes of how the transformative innovations of eco-smart energy supply chain management could be operationalized and implemented on the future Metaverse. We highlight envisioned ways of how the connected, eco-friendly and multimodal immersive innovations and models [i.e., Digital Twin

(DT), Internet of Things (IoT), Machine Learning (ML), Artificial Intelligence (AI), Virtual Reality (VR), Extended Reality (XR), Computer Vision (CV), Smart Sensing (SS), Blockchain (B) and Ethereum (E), Proof-of Work (PoW), Proof-of-Stake (PoS), Edge-and-Cloud Computing (ECD), Data Centers (DC), Communication and Functional Optimizations (CFOs), inter alia] might enable or challenge the envisioned "SmartEnergyverse" eco-supply chain management across multi-fused principles and trade-offs. These connections extend across ecological energy production, storage, customer-to-customer transmission and return on investment (ROI) across distributed energy stakeholder routes [e.g. monetization (M), "greener" knowledge performance indicators (KPIs), user (consumer) quality interface and experience (UCQI, UCQX), scalability, security, data privacy, regulatory standards, accreditation, etc.]. Chapter 3 hosts the eco-friendly virtual factories or plants (EFVFs or EFVPs) that are forecasted to support an optimized and ecological cost-efficient energy sector management in both our current energy industrial landscape and on Metaverse. We present an outlined reflection or "calibration" of the predicted Metaverse enabling intelligent technologies, solutions and human-machine energy stakeholder interaction with the envisioned eco-smart energy grid resources, the EFVFs (or EFVPs) and the connected energy channels [i.e. digital twin (DT), edge and cloud computing (ECC), artificial intelligence (AI), machine learning (ML), virtual, augmented and extended reality (VR, AR, XR), wind turbines, hydro plants, photovoltaic panels, electric vehicle stations, energy investors, founders, communities, government, housing, industrial buildings, regulatory bodies, etc.)]. Chapter 4 introduces the non-fungible tokens (NFTs) monetization for smart eco-energy contracts and sensing computing that can be applicable across the Blockchain (B) energy trading and supplier networks, framing the foundational, future "IntelligentEnergyverse" ecosystem that brings forward or aligns with the energy consumers, the energy utilities and the connected stakeholder and smart technology foundations. Finally, Chap. 5 outlines the need to generate greater scientific, research outreach, enterprise and public engagement over a future Metaverse, eco-sustainable and integrated, smart energy Grid-learning and development (L and D). Such networked learning, professional development and energy community connections could be facilitated by progressing the current scarce Metaverse, eco-smart energy supply chain management programmes in higher education (HE) and enterprises [i.e. 3-dimensional (3D) or immersive visualization systems, energy sustainable batteries, smart factories or plants, etc.)]. In that respect, future Metaverse research, development, innovation and enterprise (RDIE) would align with addressing any articulated technical, energy efficiency, affordability, safety and social governance concerns in an informed and responsible way that benefits all.

Larnaca, Cyprus — Eleni G. Makri, Ph.D.

Acknowledgement The scientific editors of Springer Nature Book Series on Studies in Smart Technologies and the publisher's production team are to be acknowledged for their invitation to collaborate and their help in the various stages of producing and developing this volume, respectively.

Competing Interests The author has no competing interests to declare that are relevant to the content of this manuscript.

Contents

About the Author

Dr. Eleni G. Makri is affiliated with Warnborough College UK/Ireland and Unicaf/Malawi/Zambia/University of East London. Her academic qualifications include a graduate degree in philosophy (of technology), pedagogics and psychology, UK postgraduate studies in psychology, industrial psychology and an awarded Ph.D. Her thesis was on social and organizational psychology. Her academic and business research and authorship publication agenda covers 47 peer-reviewed, cross-discipline and cross-cultural journal, conference, chapter, book editions, papers and presentations, including guest editorships, reviewer/TPC and COPE committee memberships and keynote talks, aligning with multimodal themes, as next. Her engagement is expanded to UQI/UQX research on Metaverse and Eco-Open Innovation, Smart Grid and Metaverse and Generative AI product development for Investor/Founder Enterprises. She also serves as a GenAI educator and instructional designer.

Chapter 1
Metaverse and Smart Grid in Energy Sector: An Open Innovation Challenge

Abstract Metaverse is considered to be a promising transdisciplinary "future ecosystem" topic for academics, researchers, industries, services, regulatory agencies and governments alike. The escalating global energy demands, power shortage, net zero emissions universal target, and climate disruption have triggered the next generation research on smart grid(s) supported by emerging technologies and renewable power resources. This chapter intends to sketch our short, envisioned outline of current, yet scarce but gradually progressing open research challenges that align with the "future operationalization" of the energy Metaverse and smart grid(s). This operationalization "mirrors" an external to internal and vice versa (i.e., open innovation) "engaged" infrastructure(s) built upon shared and digital innovation (i.e., smart) technologies and agents. "Environment-friendly" or "Green"-led energy generation, storage, distribution and consumption network, multimodal cutting-edge sensing, data analytics, fog-and-edge computing, security, governance, maintenance and monitoring denote our envisaged emergent future challenges that seem to be evident across the smart grid(s) on energy Metaverse open innovation-led future ecosystem.

Keywords Energy Metaverse · Open innovation · Smart grid(s) · Sensing · Fog-and-edge computing · Renewable resources · Safety · Screening · Governance · Decentralized · Sustainability

1.1 Introduction

Metaverse is operationalized as the future Internet. In this expanded "next intelligence" ecosystem, multimodal technologies are expected to drive and sustain multiple sensor-led interaction within extended reality (XR) environment(s). In this intertwined and shared Metaverse-XR network (space), the virtual (i.e., digital innovation) and the human (i.e., physical) technologies and entities or agents [e.g., augmented reality (AR), virtual reality (VR), citizens-clients, groups, businesses, services, government, agencies, etc.] will be synchronized and advance.

E. G. Makri, *Metaverse and Smart Grid in Energy Industry: An Open Innovation Challenge on Trends and Applications*, Studies in Smart Technologies,
https://doi.org/10.1007/978-3-032-29666-5_1

The conventional electricity grid infrastructure is considered to consist of electrical network devices that supply organizations (businesses) and residency consumers with electrical power generated from a centralized power station. Reflecting upon recent developments in information and communication systems (ICS) and emerging technologies (ET), we think that they denote powerful ecosystems which advise us how best to design, assess, store and deploy a more robust and efficient provision of "the next frontiers" power generation ecosystem.

In our current time, it is common knowledge that there have been universal concerns about climate disruption (e.g., "echoing" the goal of reduced global carbon emissions) and increasing demands for energy preservation (sustainability) after enhanced energy consumption needs. These challenges have urged the academia, research society, business, communities, policy makers, regulatory bodies, and governments to align with the demand to evolve, adopt, and maintain advanced sustainable energy technology needs and supply chains sub-systems. Within this frame of reference, we think that there has been a shift from the traditional electricity grid infrastructure (e.g., power generation, supply chain, maintenance, and distribution) to the new "intelligent", "smarter" electricity grid infrastructure. This integrated (inclusive) network is operationalized as "the smart electricity grid" or the "smart grid". It denotes the next generation of "intelligent" electricity grid ecosystem. This infrastructure appears to merge the conventional electricity grid, the emerging technologies, resources or applications (e.g., sensor metering and controlling devices, remote connections, self-restoration, balanced high power demand, lowered infrastructure costs, and enhanced client-end power generation) and the renewable resource modelling (e.g., [1]). It is envisioned to be an optimized development of best practices that ensure cost-efficient and sustainable electricity through lower emission power generation, repository, distribution (or diffusion) and deployment across businesses, plants, utilities, manufacturing, services and consumers. The end result is envisaged to be a generated ecological energy sustainable ecosystem (e.g., [2]). Using social and psychological terms, we operationalize the "smarter grid" infrastructure as the "the next frontiers", "intelligent", networked, power ecosystem that encompasses the attributes of all interrelated sub-electricity or energy systems, entities, agents, devices, and clients. We predict that this intertwined ecosystem can be optimized to be designed upon a visible, affordable, secure, scalable and power-led sustainable architecture (e.g., [3]). The following description of our "alternative future" generation modelling, insights and trends as reflected upon the sustainable energy Metaverse and empowered by smart or smarter grid(s) sensing and fog-and-edge computing technologies are outlined below, as next.

1.2 The Sustainable Metaverse-Enterprise Open Innovation "Engaged" Network

Analyzing (i.e., interpreting) the limited research performed on the future Metaverse-led ecosystems and their potential impact on the global, digital innovation (i.e., transformation) of our industrial world (e.g., [4–6]), Asian countries appear to conduct the largest amount of related transdisciplinary studies. Among these, emerging and emergent technological country pioneers serve the evolution of immersive and sustainable Metaverse solutions that can revolutionize the worldwide infrastructure across multiple sectors. Nevertheless, we reckon that any research aligned with the Metaverse envisioned continuum would need to support the detection and fine-tuning of possible challenges or concerns that might be revealed in-between the Metaverse network and the enterprise domain life-cycle.

The creation (i.e., co-design, co-development, open innovation) of the Metaverse-first digital innovations bears the potential to fuse a radical digital transformation, against the usage of physical assets and thus, benefit the sustainable power (energy) consumption. The forecasted Metaverse workspace seems to offer possible benefits in terms of envisaged client (consumer) ease of use, assets, autonomous information and communication systems (ICS), robustness, human-factor design, virtuality, visualization and safety. In that respect, any further research on the possible Metaverse-led effects on various enterprise sector(s) trends and (or) applications, seems to be foundational in order to reveal their dynamics and to foster virtual innovations. In extending the above generic operationalization of envisioned Metaverse-industrial workstation, we conceptualize the following open innovation-and-sustainably-led Metaverse-enterprise model in Fig. 1.1, as next.

This model is intertwined across three pillars, as follows. First, the human-factor client (user) experience across the whole network. Second, the long-term, robust quality of the ecosystem. Third, the sustainably supported external to internal (i.e., outflow to inflow, open innovation) transformation of data, ICS and procedures along the whole life-cycle. As regards the first pathway, it seems to be well envisaged that the Metaverse platform is expected to offer an immersive social exchange for customers (consumers). In terms of Metaverse robustness, this feature seems to rest within its dispersed network capacity (agency) that is tuned to sponsor resilience over errors or attacks. Closely related to the above, it seems to be the powerful safety that needs to be established in order to protect data privacy and unlawful access through encryption and authentication. In addition, upgrading preparedness to empower the integration of latest intelligent systems and leveraging solutions for self-recovery and ongoing screening and maintenance seem to be critical features of a future, powerful, and secure Metaverse ecosystem. Concerning the third pathway, we argue that the sustainability of the Metaverse network can be implemented through the following steps, as next. First, the greater range of virtual (i.e., remote, digital) enterprise operations performed (e.g., against physical commuting) that can decrease carbon discharges. Second, the expansion of the renewable energy resources used (e.g., wind turbines, photovoltaic panels, hydro plants, electric vehicles, etc.). Third, the

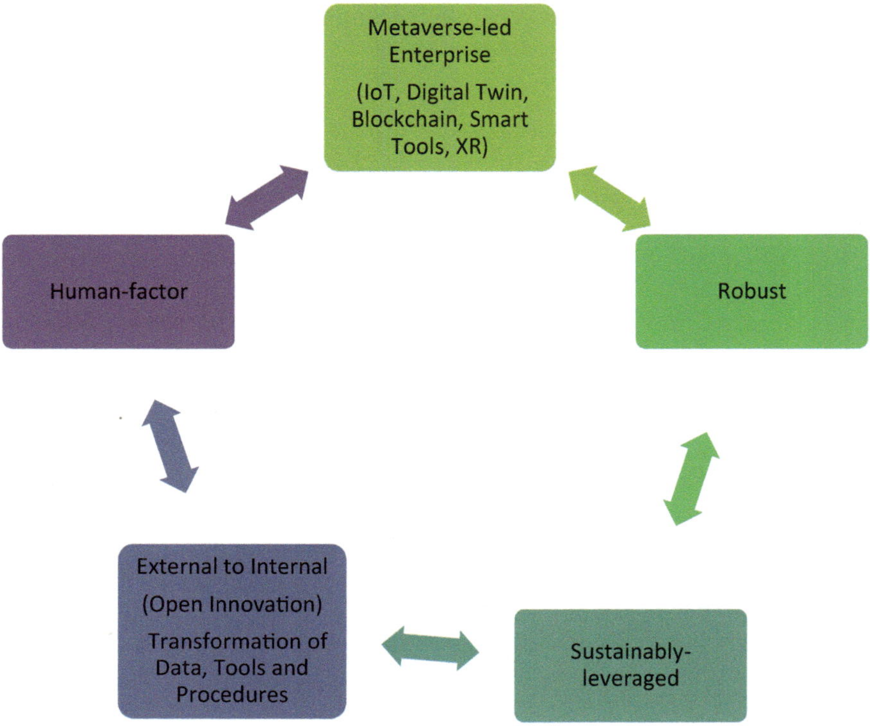

Fig. 1.1 The sustainable Metaverse-Enterprise Open Innovation continuum

application of lower-power hardware systems. Forth, the restructuring of data centers to achieve their energy effectiveness. Fifth, the expansion of reusage and decrease of waste (i.e., optimized waste management). Sixth, the ecological sustainability and resilience awareness, supported by behavioural change through education and professional practice. Seventh, the promotion of sustainability as triggered by the outflow to inflow research, innovation, enterprise, and development (R, I, E and D) (i.e., open innovation [7, 8]) exercised between environment-friendly alliances, joint ventures, enterprise networks, services, agencies, government, and communities of practice.

1.3 Metaverse and Open Challenges in Research-Smart Grid in Energy Domain

Ramírez-Herrero et al. [9] operationalize Metaverse as the future development of Web that illustrates a digital, "engaged" network where the physical and the virtual worlds are joined together. This evolving digital ecosystem is based upon diverse emerging technologies [e.g., Internet of Things (IoT), Blockchain (B), edge-and-cloud computing (ECC), Artificial Intelligence (AI), Augmented Reality (AR), Virtual Reality (VR), Mixed Reality (MR) systems] of powerful hardware, software and content features and applications. These solutions are expected to eliminate the gap between virtual tools and real-world systems by transforming our "Intelligenceverse" living. Reflecting upon different Metaverse definitions located, the authors end up describing it as "a parallel virtual world" (p. 2 above). Hovering over a number of bibliographic analyses of recent Metaverse-based research topics, research countries and emerging technological research and application trends, we think that represented authors tend to illustrate the corresponding insights, as next. First, XR, B, AI and sensors, among others, seem to be presented as foundational research themes [10]. Second, building information modelling, digital twin (DT) and governance appear as evolving research areas [11]. Third, we argue that a general-layered Metaverse "architecture" is envisioned in order to streamline open challenges for Metaverse and assist researchers and companies in introducing innovative and disruptive improvements. In that respect, Metaverse appears to be a leading technology trend, with an increased percentage of global enterprises estimated to be investing in Metaverse-associated technologies by 2030 and beyond. Nevertheless, the evolving Metaverse of device independency and shared network, seems to be still at its earlier development stages. Permeating and transforming every industrial sector in the years to come, Metaverse is expected to reach trillions in annual revenues across multimodal domains (e.g., manufacturing, higher education, workforce learning and development, defense, sports, healthcare services, smart cities, etc.). The Metaverse clients can use this multimodal open innovation network (external to internal co-creation, shared collaboration, inflow and outflow aspects of innovation implemented) to co-design (i.e., open innovate) cutting-edge engineering tools in order to exercise immersive problem solving and lend support to reduced costs and advanced effectiveness. Given that the Metaverse bears the potential to radically alter the present digital interaction practised through solutions and (or) webpages, it seems fundamental to comprehend the dynamic nature of it and offer an all-encompassing framework to improve the domain. Metaverse might garner vast popularity, but it seems that scarce or restricted research supports certain if not all domains of Metaverse, notwithstanding its principal technologies, infrastructure and impact on global responsible research, innovation and enterprise. In addressing part of the aforementioned research gaps, a deeper analysis, reflection or operationalization of the tentative bibliographic work that existed might offer an inclusive awareness of the Metaverse development. In other words, major contributions could be aligned with the following aspects. First, to capture an indicative outline of Metaverse frameworks published on the most popular global databases in order to define the core researchers, journal lists, regions and associated resources. Second, authors, academics, researchers and expert groups to be informed about the Metaverse evolving developments and core research routes linked to it. Third, to illustrate a generic Metaverse-based modelling that incorporates the major emerging solutions and infrastructure elements, as well as the open-ended challenges and future envisioned developments to support the Metaverse research and industry stakeholders by spotting inquiry spaces that need to be further explored. Overall, the three-dimensional (3D) Cyberspace of Metaverse can be envisaged to resemble the following properties, as next. First, the state of immersion, denoting the realistic experience gained by users (consumers, clients) through intelligent-based equipment. Second, the status of interoperation, with users of a unique digital identity moving between environments or sub-Metaverse areas in a seamless (not interrupted) way and interacting with the physical setting. Third, the feature of heterogeneousness across software, hardware and content devices, servers, storage solutions, interrelated networks and disperse human-machine interaction. Fourth, the element of pace and (or) development on demand, in terms of response time, following expanded use. Fifth, the state of sustainable networking by non-exhausting natural resources and environmental sustainability and resilience equilibrium (i.e., adhering to UN 2030/2050 Sustainable Development Goals for climate change, in particular). We envisage that the "Metaverse World" bears the dynamics to influence ecological, societal and financial concerns. For example, it can support lowering carbon emissions through digital avatars (or assets) replicas and restriction of physical travelling. However, security, privacy, maintenance, and no monopoly ownership and governance concerns need to be taken into critical consideration. Sixth, the long-term duration of the Metaverse intelligence and its constituent elements, as evolving condition(s). Finally, seventh, the socialization or social interaction on Metaverse which "echoes" a newer development on immersive social virtual life experience. Despite the rising popularity, the countries that have published the highest volume of Metaverse Cyberspace-led research papers appear to be in Asia, North America and Europe [12]. Our search denoting Metaverse-led notions, trends and interventions reveals Metaverse as a generic term, followed-up by diverse, emergent topics. For example, sensors, democracy, AR/VR/B technology in teaching and learning, brand marketing, aviation, gameplay, AI, media, wearable human-machine interfaces, privacy, security, and social and technological concerns, inter alia (e.g., [13]). In terms of next generation (future) Metaverse research clusters, these can be tentatively categorized, as next. Building information architecture, visualization capabilities, sustainable buildings and smart cities, DT, computational technologies, learning, monitoring, tourism, healthcare, generative artificial intelligence (GenAI), urban development, energy, ethics, societal concerns, boundaries, freedom, independence, cybersecurity and inclusion, among others [14–16].

Summarizing the Metaverse research theme categories, these can be classified as follows. First, the emerging solutions (i.e., engines and infrastructure) that embed and facilitate the Metaverse ecosystem. The engine's level includes solutions like B, DT, AI, AR, VR, XR and the ecosystem one embedding the ECC, IoT, storage, networking, and hardware applications. Second, the Metaverse infrastructure as a wide-ranging tool and an intersectoral-layered space. The infrastructure level includes the virtual assets, products or tools, and the implementation market level. In the product level, digital avatars contribute to the development of the virtual context, with offered services for virtual logistics and (or) e-commerce. The infrastructure level categorizes solutions into digital tools and objectives. The first reflect challenging domains like teaching and learning and health services, while the latter illustrate the Metaverse agency for gaming and enterprise developments and productivity.

The intersectoral tier-space spans across the Metaverse ecosystem and addresses features of security, control, privacy and social interface. Finally, the emphasis is shifted towards the user-centred nature of Metaverse, signifying the connected open-ended strands, research and future insights on Metaverse, as follows. The first cluster reflects the hardware ecosystem as the sectorial physical environment (i.e., ECC, storage, networking, IoT, and sensor tools) that embeds the users' actions and devices (e.g., haptic, sensors, etc.). The transmitted data are progressed through the ecosystem and should accomplish low reaction time (i.e., latency) and trustworthiness to facilitate the immersive nature of the Metaverse. The networking technologies like 6G and (or) IoT strengthen the recurrent ecosystem accessibility and real-time gigantic data diffusion. However, the advanced cellular wide-bandwidth of enhanced trustworthiness and low response time ICTs tend rather not to be yet so well-crafted emerging ones. The endorsement of multimodal varied ecosystems (e.g., including wireless broadband and satellite systems) seems to denote a core Metaverse research avenue. Offering a user-centred, engaged user experience by capturing the full sense of worldwide sensor and seamless connection, appears to be of utmost importance. In addition, the deployment of energy-effective ICT modus operandi and the implementation of energy-reaping methods for sustainable ecosystem function and ecological networking should become a major future research direction. Moving forward but reflecting the aforementioned properties and classification of the Metaverse "alternative future" research domains, Figs. 1.2 and 1.3 replicate an outline of the "next generation" (or "next frontiers") researched ecosystem of the Smart Grid on the Metaverse that energizes the Energy Sector, in particular.

As regards the Metaverse technology ecosystem of ECC, we reflect upon published explorations (e.g., [17, 18]), as next. First, the ECC devices and services supported by an enhanced mobile broadband are able to facilitate XR, AI, DT and B technologies, resting on supported hardware ecosystem. In addition, they can facilitate the tools' performance, their scalability, their computational (smart) intelligence and strong storage agency, advancing human-machine interfaces. ECC appropriate resource allocation renders a great challenge, given the advanced requests with extended usage. In addition, the option of "greener" edge computing (e.g., [19]) is presented as necessary to develop inventive methods that foster ongoing computation service. Blockchain technology is expected to advance the Metaverse with newer consensus algorithm protocols (e.g., [20]) tasked to ensure trustworthiness, security, scalability, and decentralized monitoring. These protocols need to be energy-efficient to prevent negative impact on the physical environment (i.e., gas emissions). Further, advanced interconnection(s) should be facilitated to support the improvement of interoperable and decentralized Metaverse sub-chains efficiently, as empowered by energy-friendly blueprints. In addition, diverse meta-machine learning approaches should be expected to be available on the Metaverse to address the auto-ML challenges employed, with GenAI gradually transforming the information and communication paradigm across the Metaverse platform, as well. Despite considerations over the future developments on an "integrated" (or "inclusive") immersive Metaverse, yet, various challenges that denote ethical usage, data privacy, security, and maintenance tend to become key drivers for meticulous reflections. Further, advanced

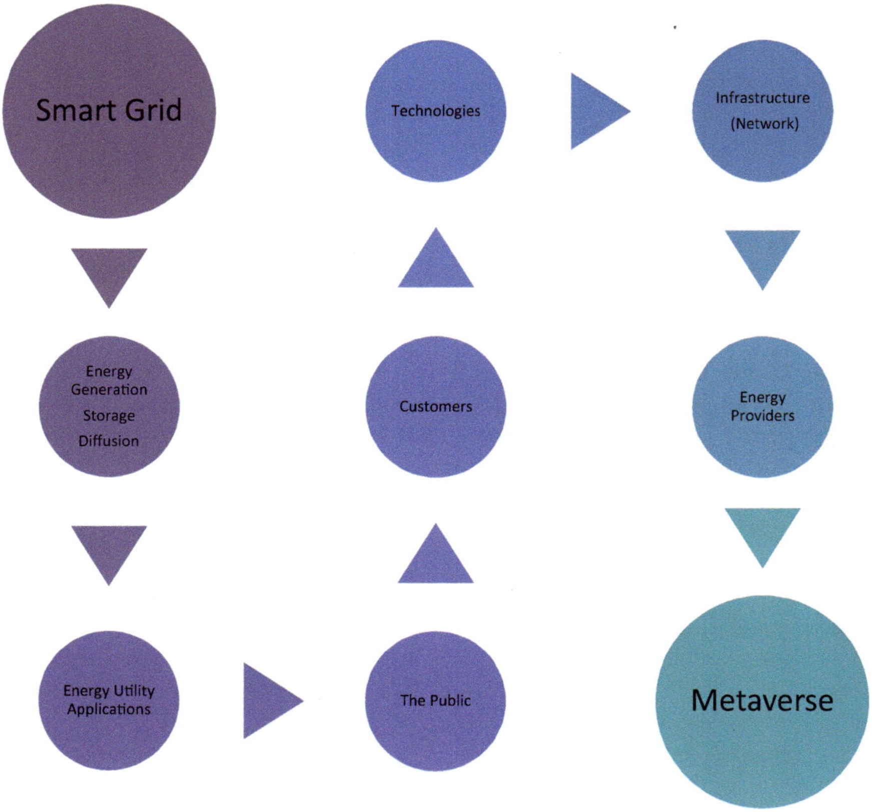

Fig. 1.2 The smart grid and the Metaverse

sensory modalities need to be fully integrated into visual Metaverse immersion with affordable cost-efficiency to support greater virtual adoption. Additionally, the digital Metaverse configuration along with authentication access screening seem to be still at a rather early stage of development, with the present state devoted to user-machine interaction-driven techniques (i.e., representing data inflow fusion). Based on the aforementioned evidence, rationale and reasoning, we might envisage that the development of Metaverse-enabling solutions is still rather early enough. However, we argue that it is expected to boost on top scale in the following years. The performed bibliographic studies revealed transdisciplinary Metaverse-led domains, concepts, interests, and expertise fields. We conclude that it tends to be vital to embed and enhance multimodal solutions on the Metaverse ecosystem. Knowledge transfer from innovative businesses that rely on content, product or service design (co-design) and deployment along with accessibility and privacy resolutions on virtual environment(s) needs to gain deeper consideration. Overall, we need to emphasize the existence of open research challenges and future directions in Metaverse cross-over fields, customers, interventions and emerging technologies.

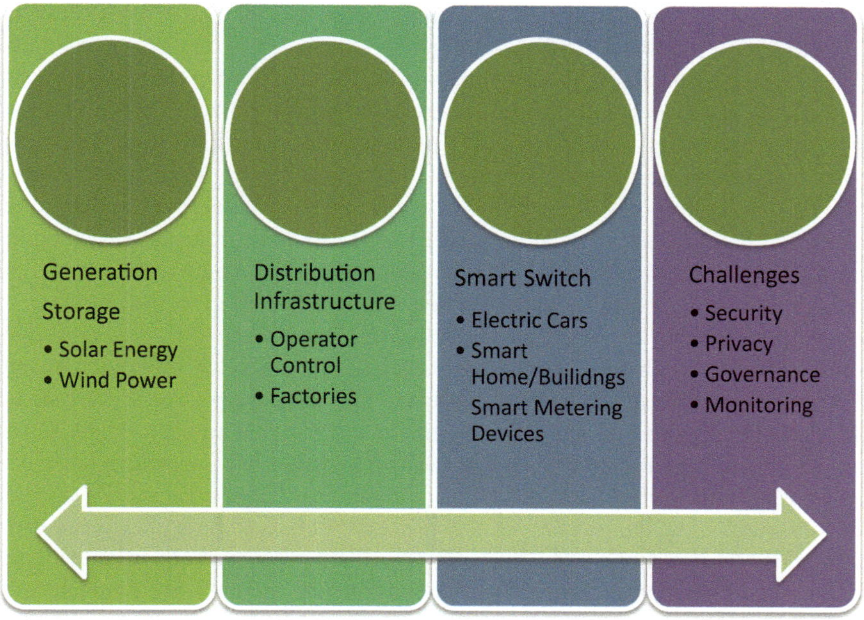

Fig. 1.3 The smart grid on Metaverse that fuels the energy sector

1.4 Energy Metaverse and Smart Grid(s)

Tightiz et al. [21] and Fan et al. [22] review and elaborate on the research reflecting the status of energy-promoted Metaverse and Smart Grid as the "alternative futures" or "next frontiers" energy structures, resting on their expected smart intelligence, digital and information-fostered nature, and dynamics. The basis of their exploration tends to span across the power-based input attainment through extensive dispersed detectors (i.e., sensing). These "sensors" (e.g., [23, p. 1]) seem to be multi-criterion, minimized, of great accuracy, autonomous in power, wirelessly transmitted, and highly trustworthy. The authors reflect on the considerable developments in advanced, operation-led, autonomous-powered, small-scale sensors that have been showcased through the multimodal integration of nanotech and micro-electromechanical detectors (e.g., p. 4 above). However, they emphasize the fact that despite the various reports of autonomous sensors in commercial environments like smart healthcare, smart urban design, development and farming, there seems to be rather limited overviews for energy grid(s) [21, p. 1]. In this respect, therefore, they review the latest developments and applied implementation and outcomes of autonomous, multimodal and micro-sensing digital innovations for energy (power) Metaverse and Smart Grid. In particular, they meticulously discuss the cutting-edge developments in micro-sensors, the power-based leveraging strategies appropriate for energy grid(s) and the implementation of the autonomous power-led, micro-detection in power grid(s). They move forward by considering the challenging issues and the future work prospects in co-designing the Smart Grid by building upon the autonomous, power-led multimodal sensing technologies. In more practical terms, they further discuss the worldwide energy deficiency and the climate disruption, with the zero carbon emission global goal becoming a shared need (or vision) among universal regions. Advancing the utilization, for example, of renewable energy resources (e.g., wind turbines, solar panels, hydro plants, electric vehicle stations, etc.) to address (moderate) the worldwide energy shortage and empower the global environmental sustainability and resilience. Such interventions can be achieved by co-designing affordable and energy-sustainable power networks that are able to decrease carbon emissions, not only in industrial but also in domestic sectors. Inclusively, the circulated renewable power of the grid-link and the implementation of electrification seem to constitute the energy grid infrastructure more dispersed and the energy stream multimodal. This condition tends to decrease the fine-tuning capacity and impacts on the steadiness of the energy structure. As an outline, the energy grid(s) appear to exercise a critical role in the diffusion of the electric power, the shared energy provision and the substitute energy allocation, but they also showcase a core agency in ensuring a stable and trustworthy electric power supply chain continuum. In an operationalized sense, the smart grid(s) is (are) built upon physical energy, embed sophisticated sensing, metrics and control tools to accomplish calibrated power distribution and device(s) position, improving their capacity to locate errors, integrating risk detection, and ensuring suitable safety that advances reliability. Reflecting on the above authors' insights, we sketch their envisioned smart grid structure that incorporates the Metaverse notion in Fig. 1.4, as next.

Following up on the above power Metaverse and smart grid system operationalization, it is evident that Metaverse appears to be the final frontier of digital-first innovations that advance the information and communication intelligence of the energy network, reflecting a comprehensive and synchronized interaction of the physical

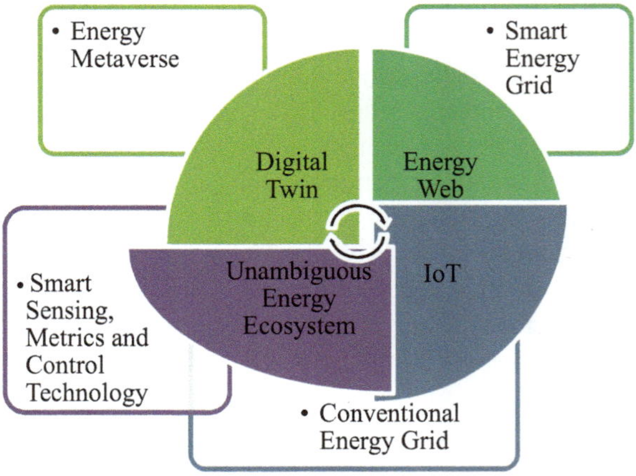

Fig. 1.4 Energy Metaverse and smart grid network operationalization

energy grid(s) with the virtual world. In this projected interface, the energy Metaverse is expected to implement diverse emerging technologies (tools) such as the IoT, DT, edge, fog and cloud computing (EFCC), big data (BD), B, AI, and other sophisticated human-machine application programming interfaces (APIs). These systems are deemed necessary to realize the virtual and physical amalgamation of human agents and the energy Metaverse, creating a smart (intelligent), human-centred energy ecosystem. For instance, the IoT technology links the physical energy grid and energy Metaverse through immense and wireless energy grid sensors. The rest of the advanced technologies, that is, the DT, the EFCC, the BD, the B, the AI, and human-machine interfaces are expected to advance the data harmonization, assessment, safety, and user experience (UX) on the energy Metaverse. In that respect, following up on relevant empirical premises (e.g., [23]), we think that the key driver to accomplish the energy Metaverse and smart grid ecosystem(s) is to endorse immersive, "eco-friendly" energy data innovations (i.e., "ecological" energy generation, diffusion, conversion, circulation and expenditure), and sensing technology. This condition "mirrors" an open-and-eco-innovation (i.e., outflow to inflow innovation resting on environment-friendly energy resources) energy (power) integrated infrastructure, as envisioned in the following Fig. 1.5.

Further reflecting on the aforementioned authors (i.e., [21–23], we think that they additionally stress the need to deploy larger numbers of diffused sensors not only for forecasting or warning preparation(s), but for garnering both power-related and environmentally based categories of data, in order to acquire more sophisticated information for the evolution of the energy Metaverse and smart grid(s). Their insights enlighten us on the variations of power knowledge performance indicators (KPIs) like electricity, water flow and energy bringing power breakdown, equipment failure, and

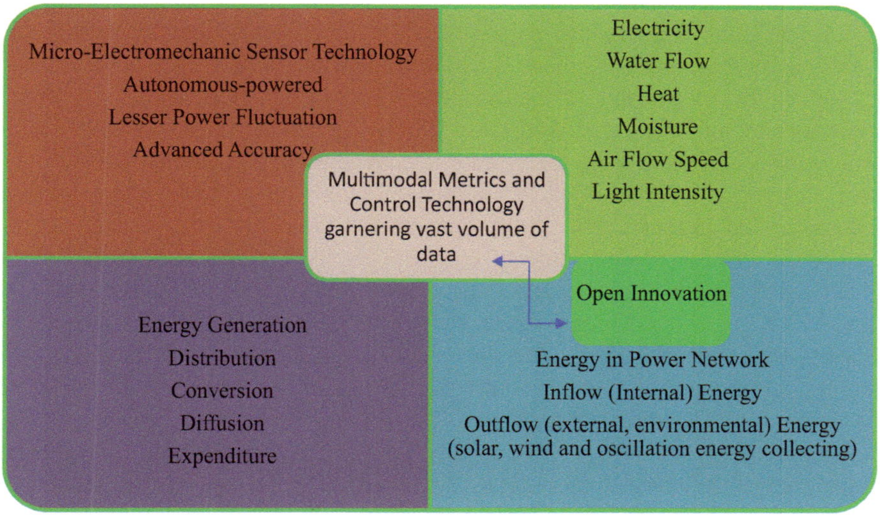

Fig. 1.5 Multimodal sensing requisites and possible power resources in energy grid network(s)

(or) safety risk. In another vein, they direct us towards considering that the regular performance of the equipment seems to be unavoidably influenced by the nearby settings. Environmental screening like heat, moisture, air flow speed and oscillation energy leveraging seem to be able to offer richer data to ensure the standard performance of the equipment. Resting on the above, we argue that it tends to be inferred that sensing devices captured as smart, multimodal, optimized, of increased accuracy, lesser energy fluctuation, autonomous-powered, wireless transmitted, of decreased expenditure and advanced trustworthiness, might constitute them rather more appropriate for diffused (i.e., distributed) energy grid(s). Aligning with author arguments, we agree that if we bring together the multimodal-boundary sensors with the power-leveraging devices, then we might be able to perform a long-standing screening of energy network(s), free of upkeeping. Notwithstanding the need for growing research on autonomous sensing (i.e., sensor) technology in domains like healthcare, smart cities and (or) smart agriculture, among others, we agree with the authors that there seems to be still, rather limited or scarce empirical evidence, reviews or meta-reviews on connected, sensing innovations for energy Metaverse and smart grid(s). Going multiple steps further, we cannot align more with their rationale that the new energy network has to assume expanded and long-term power diffusion (i.e, distribution) with elevated capacity (i.e., properties), given the power scarcity and unbalanced allocation. As the incorporation of renewable power resources like photovoltaic panels and wind turbine energy improves, the energy grid is overwhelmed by electricity fluctuation challenges, while the energy generation faces electricity issues, driving to device(s) failure, energy grid(s) unsteadiness and creating serious challenges [24]. In addition, the climate disruption, evident to all, through acute weather conditions and disasters, seems to threaten the newer energy production, with the physical environment affecting the energy load ability of any elevated-electricity transmission [21–23].

In order to tackle these challenges, we tally with the above authors arguing that lesser-energy fluctuation, micro-sensing innovations might be applied across a large domain to garner electrical and environmental data (e.g., electrical energy, wind, moisture, among others, etc.). Following climate disruption vast progression, excessive weather conditions take place more often, ending up to electric energy equipment failure and disorder of energy provision (among other breakdowns). Renewable power-led dispersed generation properties, like the solar panel and wind turbine energy are sensitive to environmental elements, risking irregularity [25]. In order to guarantee the safety, trustworthiness, and sustainable performance of energy grid(s), developing micro-digital innovations for real-timeline climate screening systems customized across diverse physical and weather conditions, tends to be critical for forecasting future grid(s) function, ensuring well-timed warning of possible climate risks and preventing disasters (as above). Further, the prior reported authors [21–24] emphasize the significance of real-time screening of the electrical and environmental boundaries of the energy grid(s) network, in order to achieve an optimal functional status. By doing so, an early discovery of possible challenges and unusual conditions might allow for proactive adjustments and appropriate maintenance. This would ensure the safety and stability of the energy grid(s) and could offer core metrics for the energy Metaverse and smart grid(s), as well. We associate with the aforementioned insights and additional ones of the same authors (i.e., [21–24]) postulating that the development of the energy Metaverse and smart grid(s) seems to be an advanced sophisticated engineering venture that assumes a vast amount of sensors. The latter tend to get into the overall procedure of energy production, diffusion, conversion, allocation, and expenditure, offering a newer type of cross-over and multimodal domain implementation. However, we could not agree more with the authors' suggestions that there are remaining challenges to be resolved. We characterize them as open innovation-based, outflow and inflow, systems dynamic modelling issues that appear imperative to be addressed, as next. First, the nano-engineering needs to accomplish advanced accuracy, extended range and flexible sensing innovations for multi-level boundaries. Second, the energy leveraging technology has to develop energy resources with elevated energy solidity, stability, duration, and incorporating a system-based implementation and manufacturing technology to accomplish the optimization of the sensing innovations. Third, to harvest the edge-intelligence technology to establish the configuration of a cloud-edge, shared energy grid(s) function. Fourth, to progress on newer wireless communication that serves the multiple-source data transmission, ensuring privacy and effective resource allocation. Fifth, expanding the DT technology to manage the precise replicas in virtual energy grid(s), illustrating the status, connections and functional traits of the physical energy grid(s). Sixth, to leverage the cloud computing technology in order to explore, calculate, and analyse robust methods for expanded data extraction and for greater energy network flow, consistency investigation and security evaluation, developing innovation robustness. Seventh and final, to focus on emerging technologies like XR systems to obtain immersive, operator-electrical device interfaces, advancing the data visualization of energy grid(s) networks.

1.5 Technology-Enabled-and-Primary Innovations for Net Zero Emission(s) Transferred to Energy Sector and Metaverse

Ahmad et al. [26] and Agrawal [27] elaborate on the persistent digital and (or) immersive conversion along diverse aspects of our life, yet emphasize the enduring physical risk of global climate disruption. They claim that accomplishing or moving towards energy generation of net-zero emission(s) across industrial sectors [26, p. 1], might well address the optimal goal of decarbonizing our world. In that respect, digital, virtual or intelligent innovations could be the key drivers to speed up evolution on net-zero emissions, based on their prospective intelligent capacity to assess, control, manage, and alter the emissions reality of our current economic development. Stern and Valero [28] illustrate a newer overview on technology-driven innovation that assumes eco-sustainability, by investigating the ways which fast developing automated innovations might support net-zero properties across diverse domains. Cao et al. [29], Falcke et al. [30], Uchenna Attah et al. [31] and Virmani et al. [32], among others, outline the use of four generic, smart net-zero methods by integrating and extending previous related work. Their operationalization(s) can be used as a triggering spot for current and future directions and inquiry, offering potential roadmap(s) on strategies that assume net-zero operation. Their envisaged target is to challenge current research on digital sustainability exploration and exercise, by proposing to back up emissions decrease through removing and regenerating them by using technology-led innovation(s). Along this line, businesses might well design their net-zero organizational operations following up on the principles of France's global agreement on net-zero carbon emissions. Yet, it is evident that corresponding compliance with net-zero strategic focus is still very low. On one hand, governments can shape the enterprise response to climate disruption and motivate decarbonization when they do not seem to comply with ecological regulatory standards. On the other hand, interested communities can put a real pressure on leveraging carbon decrease targets and progress. Nevertheless, we suggest that a combined entrepreneurship (i.e., open innovation) overview might be beneficial to address decarbonization goal not from an organizational standpoint per se, but from a digital (immersive) systems design thinking (i.e., open innovation) to achieve sustainable strategy through the development of prospective business models with ecological wins. Engaging external and internal stakeholders in outflow to inflow exercise of innovation (i.e., open innovation) could transform the emission-powered businesses closer to net-zero ones (e.g., using renewable networks). Despite the large-scale carbon emissions reduction needed, various high-tech companies have invested billions of dollars to calibrate carbon emissions elimination. Yet, removal and regeneration remain corner topics across the related literature. Technology-driven innovations can enable the control and optimization of emissions-fused procedures through digital data interaction. In addition, carbon deduction entails well-operating carbon marketplaces which rely on the digital illustration, authentication, and trading of carbon elimination certification (accreditation).

We align with the above authors [28–31] in that "technology-led innovations" seem to reflect the use of digital assets to illustrate the optimization of physical ones, while the "smart or intelligent or cutting-edge or technology-primary innovations" tend to denote employing digital assets to re-create interventions in the physical environment (i.e., on-site). Exploring both strategies, it might be helpful to investigate different modes of harnessing technology-aligned innovations in favour of net-zero target. In that sense, it might be advisable first to create technology-supported net-zero designs (systems) and then to proceed with smart-first (technology-primary) net-zero plans.

Technology-Supported Net-Zero Methods

The above authors [28–31] aim to propose newer insights or strategies on technology net zero-enabled innovation. They first synopsize previous inquiry and spot critical puzzles associated with net-zero management paradigm. Next, they elaborate on sustainable tech-savvy innovation-related work to support addressing the aforementioned issues. The following parts describe their explored technology-facilitated net-zero tactics, where businesses employ technology innovations to signify and promote physical resources and value-added chain management to alleviate discharges and re-develop.

Technology-Fostered Elimination and Emission Decrease Approach

The first method entails technology-supported innovations for emission moderation and diminishment. According to their insights, indicative examples of fundamental issues that promote emission decrease, moderation or calibration consist of the following steps, as next [30, 33]. First, the assessment of target-linked context and discharges. Second, the implementation of the needed pace of lowered-carbon innovations. Third, the alignment of company frameworks with discharges restriction. The above challenges tend to be rather enduring ones, given the business divergent functions and the incomplete regulations and optimal trials needed to spread accountability for carbon discharges along the supply chain cycle across diverse sectors. In addition, current enterprises have to address existing technological disruption (i.e., which solution fits best with which application system) and (or) multiple choices for decarbonizing. Further, in cases where the technology option is clear-cut, a critical dimension seems to be the rate of endorsement, particularly within diverse companies that need to control their innovations. Lastly, various business entities tend to experience the puzzle of lining up the ecological and cost-effective merit in their enterprise strategy. As a starting point, the above authors argue that the promising elimination of discharges seems to demand advanced ICT that supports companies to garner, stockpile, and handle emission(s) analytics (e.g., providing physical networks with sensing capacity). The Cyberspace (including the "alternative future" Metaverse) tends to allow for and (or) reveals newer ways to transmit shared metrics on emission(s) and keep business responsible for their actions. Moreover, the technology-advanced or intelligent innovations are able to upgrade discharge-demanding (physical) infrastructure, procedures and actions in favour of sustainability maintenance (e.g., through enhanced energy effectiveness and optimization of manufacturing procedures). In addition, they can support and (or) motivate towards a more energy-efficient human behaviour, manage diverse agents (i.e., stakeholders) and (or) assets for emission(s) control and introduce supplementary lowered-carbon uses and company modelling. For example, in energy technology businesses, sensor solutions and (or) sophisticated data analytics can allow for the assessment of wind turbine pathways or smart intelligence devices (i.e., drones) might garner metrics for developing precise digital replicas of physical wind pathways, among similar networks [34]. Such digital replicas are linked to advanced databases that are able to assist operators to be vigilant in immediate action when notified for recurrent outages in specific cases, with the workforce acting in response, by reimbursing physical resources and networks.

Cross-Over Related Work on Energy Net-Zero-Associated Puzzles

Continuing our reflective elaboration on the above authors' rationale and reasoning, it seems that smart controlling through technology-enabled innovations is able to extensively enhance the produced renewable energy and facilitate supplementary, technology-fused and lowered-carbon utilities. Relevant example cases might entail the energy effectiveness and management systems such as the energy control sensing and software development tools introduced from energy technology companies [32]. For instance, co-designed (i.e., open innovation) applications that can assess and accurately replicate the power (energy) intervention(s) in business or housing infrastructure, allowing for optimized usage. Along this vein, the clients (i.e., end users, consumers, suppliers) that endorse these cutting-edge utilities can advance the effectiveness of their processes, reduce expenses and eliminate emissions. Resting on the above, the proposed technology-enabled innovations of net-zero (zero carbon emission) strategy can be envisioned, as next.

Method 1. Technology-driven emission alleviation by employing advanced data science tools to replicate and optimize the physical processes and design supplementary utilities for the restriction-targeted interventions [28]. Adopting this approach as a preliminary stage, a call is extended to explore the contexts, antecedents and (or) outcomes under which enterprises can harness technology-led innovations to improve or supplement the physical resources supply chains with digital agencies and utilities to restrict emission(s).

Technology-Supported Exclusion and Renewal Method

The following part illustrates indicative ways in which the technology-enabled innovations can further address management puzzles that relate to carbon emissions deduction and nature revival (e.g., [35–37]) as next. Following the above authors' reflections for future development, carbon exclusion industries need to be developed radically and with accelerating pace. However, current digital innovations seem to be expensive and unscalable. Therefore, some pervasive questions have to come forward, as next. For example, how can enterprises efficiently manage their research and development (R&D) to support the design (co-design) of innovative and cost-effective carbon elimination and renewal systems? How can businesses synchronize their R&D and commercialization attempts to size carbon exclusion and revival innovations? Given that diverse tools (systems) are rising, it appears to be demanding to validate the quality of carbon deduction and restoration attempts, particularly when referring to their contribution and longevity. In addition, enterprises tend to experience the riddle of developing sustainable marketplaces where authorized members can rely on the quality of carbon exclusion and renewal. Last but not least, the carbon elimination and physical restoration signifies a regulatory organizational model innovation. How can companies harness renewal ethics to create and balance organizational models (i.e., frameworks) that exert not only merit by promoting carbon deduction validation but also permit the creation of innovative goods and utilities? For instance, some enterprises have experimented with seizing carbon as a component in revival farming. Although carbon exclusion seems to be naturally physical, technology-led innovations bear the potential to support speedy technological development, cost-effectiveness and harmonizing request and provision.

We think that the aforementioned operationalizations and envisioned author insights seem to lend support to the narratives that related work currently offers rather limited data into how technology-enabled innovations can speed up carbon deduction and nature restoration. In going forward with the above author's further premises [28, 34–36], we align with them in stressing the importance of sustainable, authorized and (or) regulated, carbon-decrease digital innovations. For instance, an energy innovation company can employ sensing technology to improve its physical wind turbines. While carbon alleviation seems to be a physical function, the marketplace might only operate if regulated, digital replicas of sustainable carbon deductions survive. In this respect, technology-enabled innovations tend to be critical for levelling up the carbon exclusion business. The second author-suggested net-zero promoted method can be envisaged, as follows.

Method 2. Technology-led discharges exclusion and renewal approach by harnessing digital replicas and data analytics to precisely validate and improve physical-driven carbon alleviation solutions and control revival attempts [28, 30].

Adopting this approach as an introductory stage, we reflect upon the aforementioned authors' related queries seemingly coming into effect, as next. First, how can companies harness technology-enabled innovations to manage carbon exclusion as a revenue intervention option? Second, how can enterprises use technology-led innovations to improve physical R&D procedures on carbon alleviation? How can organizational entities optimize IoT solutions to replicate an advanced, physical carbon elimination procedure and design (e.g., co-design, open innovation) a data analytics-driven company strategic intent? Beginning from the physical procedures of carbon exclusion and renewal and considering technology-supported innovations as replicas, optimizers and facilitators of this physical operation, seems to be a core research aspect that needs to get more attention in business and management disciplines.

The two prior technology-enabled net-zero methods synopsizing insights from related work on computational-led, eco-friendly paradigm from diverse areas, tend to offer functional information through relevant examples and showcase future inquiry options associated with net-zero framework. Provided that climate disruption tends to be a rather physical risk linked to the discharges of current physical networks, further creating technology-fostered methods for both discharges alleviation and revival, seems to be a fundamental action for computational-driven ecology academics, researchers and industry experts. Having said that, the technology-promoted net-zero methods might be restricted by current physical networks and limited upgrade of physical resources. Therefore, in the next section, we elaborate on how the preceding authors [28, 30] operationalize the smart (i.e., intelligent, cutting-edge, disruptive) methods for net-zero and their envisioned dynamics to support technology-enabled innovations.

Smart Net-Zero Methods

The dynamics of smart net-zero strategies as those supplementary to technology-led net-zero are reflected in work exploring emerging solutions and ICT-based innovation procedures. When adopting intelligent net-zero methods, businesses tend to begin by innovating with computational forecasting that can harness learning and development and with renewal impact by substituting and restoring physical technologies for exclusion and revival (as above). AI-led insights or frameworks, in particular, seem to demonstrate the ways learning and development can advance the assessment of sustainability and the forecasting of discharges resting on diverse data analytics tools (e.g., energy supply chain management in city context) [36]. Along this continuum, the digital spaces are able to support infrastructure and possibly address trustworthiness challenges in moderating climate disruption. In that respect, computational sustainability inquiry began to explore the generic implementation of immersive solutions such as AI, XR (Metaverse) systems and (or) digitally enabled spaces and their capacity to support sustainability.

In an outline, a smart net-zero method tends to bear less discharges, as it seems to rely less on physical resources and functions for experimentation. As a substitute, the discharges of intelligent-led experimenting originate from the energy (power) needed for core data centre networks and digital agency usually offered by big network companies. The dynamics for cutting-edge innovations to spread innovation and revenue with less extra discharges can be in place, if we can decarburize the existing computational network through renewable energy (power) implementation and developments in ICT systems. Taken together, we think that the smart-mindset tends to offer a rather innovative approach for re-shaping the business computational sustainability frameworks, by considering that harnessing digital forecasting with learning and development, infrastructure and renewal impact can facilitate emission moderation and decrease (e.g., energy industry, etc.). The aforementioned systems employ prior and actual data analytics from diverse resources, like climate, attitudinal and smart metering to predict energy generation and request. Given that these applications harness learning, development and evolution and advance by time, they offer precise computational forecasting of energy product development and need. Assuming lower cost of API development, these digital forecasting tools of energy generation and usage might be employed to design (i.e., co-design, open innovation) computational simulations and improve them through digital experimentation and learning and development impact. Resting on practice with physical networks, certain industry smart forecasting and simulations might be able to restore important parts of their physical interventions.

Likewise, in advanced-accuracy manufacturing production like automotive industry, companies gradually employ accurate digital models and ML to conduct iterative interventions only computationally, re-creating the whole procedure from physical-led to smart one with decreased discharges [38]. This intelligent mindset might restrict pathway(s) dependence and release reproductiveness, as the introduction of the innovation procedure can be to some degree separated from the physical network and related emissions (as above). In another example, when the forecasted energy production is restricted but the forecasted energy request is enhanced, clients might be prompted to use other resources (e.g., stockpiled power) and end users to spend lowered energy, harnessing infrastructure [30]. In sum, as outlined above, re-considering digital innovation from a cutting-edge, primary aspect and harnessing learning and development, restorative and infrastructure impact might help alleviating and decreasing discharges. The third net-zero method is elaborated next.

Method 3. Intelligent-driven (technology-primary) revival and deduction approach harness cutting-edge forecasts with learning and development, infrastructure and restorative outcomes to substitute and recreate physical procedures and smart utilities for discharge(s) exclusion and decrease. In order to enhance our shared reflection on the impact of this method, cross-boundary context and arrangements, the aforementioned authors [28–31] propose future inquiry to explore the ways, reasons and circumstances where smart-led innovation(s) might speed up discharge(s) decrease. In the following space, we outline their envisioned insights on the second aspect of net-zero and reflect on their dynamics of smart-driven novelty to support carbon exclusion and renewal.

Intelligent-Led Elimination and Restoration Method

Given the originality of both approaches, there seems to be rather scarce previous work at the meeting point of smart-driven innovation(s) and carbon exclusion and revival. Nonetheless, up-to-date work on digital (computational) sustainability appears to showcase the significance of immersive solutions for resolving trustworthiness challenges connected to sustainability. Resting upon this paradigm, we argue that the preceding authors might intend to trigger current research by illuminating a smarter-led method for carbon exclusion and renewal. In order to describe the notion of digital-driven innovation(s) for carbon alleviation and revival, the prior authors [28–31] reflect on applications where the company(ies) supplemented physical-primary and digitally driven solutions with intelligent-primary innovations (e.g., agriculture). At the beginning, for example, the companies created a tree-flow chart that offers a synopsis of tree-plants and replantation initiatives. The chart was at first created to be a digital replica of the physical environment that could be harnessed to advance physical procedures and design supplementary digital utilities. Adopting this as a point of reference, they integrated a digital-primary design and created the effects chart with tokens as digital-primary forecasts of restorative interventions. Their revival initiatives, cost and investment seem to appear for the first time in the digital context. The physical restoration comes next and tends to be innately connected to the price of the digital asset (coin). The platform model allows learning and development, infrastructure and production, bearing the possibility to extensively speed up the renewal of the natural infrastructure. Lastly, by mixing learning impact from a business carbon solution and infrastructure outcomes from a carbon exclusion buying and selling tool, another energy technology company's embedded platform turned to be a case study of how a digital-primary method can support the facilitation of an improved quality carbon alleviation marketplace. In adopting the above insights and logic, the preceding authors [28–31] envision the next method, as follows.

Method 4. Smart-led (technology-primary) exclusion and renewal method merges digital forecasts with productive, learning and development and infrastructure outcomes to create, prompt and legitimize discharge(s) exclusion and revival.

Adopting this method as a point of reference and reflecting on above authors' inferred insights, we align with their call for future directions exploring the ways, justification and contexts under which the intelligent-primary mindset could allow the design (i.e., co-design, open innovation) of enterprise models that blend the design and scalability of carbon exclusion and restoration with financial return on investment (ROI) development.

Going forward and integrating the four net-zero methods outlined above, one might envision a greater net-zero ROI evolution. However, the aforementioned authors [28–31] appear to conclude that the digital-led discharge(s) decrease method seems to be the most promoted across research and practice. Resting on the premise that the discharges need to be radically alleviated from our physical networks, this seems to be a rather critical point that calls for further research along this direction. As regards the rest of the three methods, the prior authors [28–31] seem to propose a similar call requesting greater investigation of the ways and associated outcomes digital-driven innovations might exercise towards carbon elimination and renewal, with the support from a worldwide net-zero financial ecosystem. Complementing this argument, we claim that they call for additional exploration of the dynamics that intelligent-led innovations might bear for alleviation and decline. As a final inquiry route, we envisage that they support exploring the combination of the smart-led innovations with carbon exclusion and restoration, as core and intertwined elements of digitally driven net-zero methods. Nevertheless, the particulars of potential business barriers, industry sector properties, stakeholder objectives and (or) restricted innovation capacity (or agency) might cause challenges or inconveniences in selecting specific or blended, digital-led net-zero methods. Provided that the net-zero notion proposes an emphasis on diminishing discharges as fast and as largely as feasible but also reveals the significance of perceptively developing a carbon exclusion and restoration business, where companies might have to purposefully consider the timeframe and the range of resources allocated towards alleviation and decrease, versus (vs) exclusion and renewal. A similar account of future exploration could be envisioned for digital-driven and intelligent-led net-zero methods. For instance, future research directions might entail an investigation of wins and (or) barriers regarding the agency of inflow business design options, such as blending digital, intelligent (smart) and sustainability operations. Also, an examination of the open innovation (i.e., outflow to inflow innovation) partnerships in sharing digital-driven and smart-led net-zero methods. As a last note, the provision and accessibility of extended renewable energy might efficiently affect the success of digital-led rigorous net-zero methods. Therefore, under context-specific circumstances and for certain companies, it might be most efficient or sustainable to select one and (or) merge multiple approaches. In an inclusive (integrated) reflective outline, our previous encompassed reflections on prior authors and own insights, rationale and reasoning are sketched in the following Figs. 1.6, 1.7, 1.8, 1.9 and 1.10, as next. In specific, Fig. 1.6 sketches the operationalized open innovation (i.e., external to internal) methods and related challenging issues for net-zero business optimization. Figure 1.7 outlines the four digital-driven and intelligent-led (technology-primary) sustainability methods for net-zero achievement in the energy area. It tends to be rather reasonable to operationalize that these four methods (or strategies) might be envisioned to be transferred to Metaverse-led energy-driven ecosystem (Fig. 1.8). Figure 1.9 outlines the comprehensive operationalized model that sketches the technology-driven net-zero methods

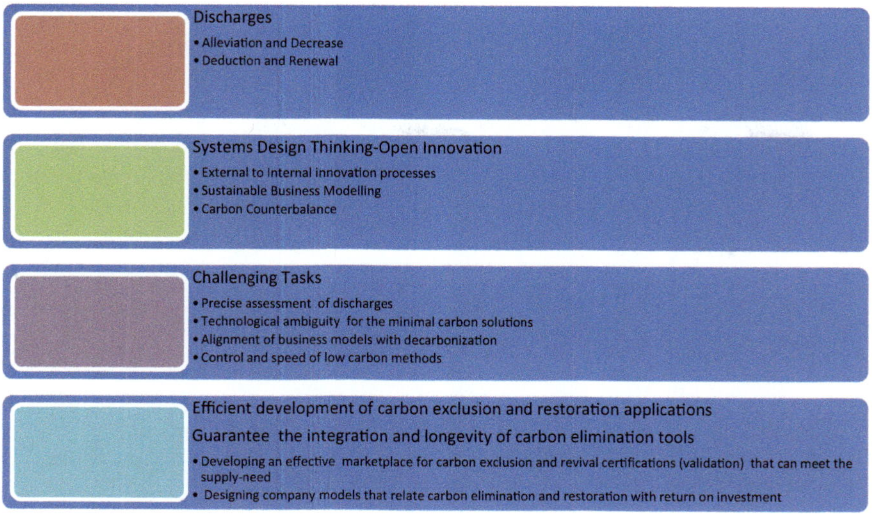

Fig. 1.6 Net-zero generic cross-over methods and related challenging issues

in the energy zone, an envisioned replication (replica) of which on the Metaverse platform is presented in Fig. 1.10, respectively.

In concluding the prior authors and own insights, the long-standing call for further research, evidence and interventions that explore climate disruption and net-zero transition continuum with sustainable business and management by distinguishing between emission(s) alleviation and decrease and exclusion and renewal, the two sides of the inclusive net-zero continuum, come into place.

1.6 A Secure Smart Grid(s) Operationalization on Energy Metaverse

Tariq et al. [39] indicate that the smart grid tends to depend on the Improved Metering Network (IMN) innovation, a technology that has been showcased to be vulnerable to invasions. To efficiently control and trigger the alarm when irregular tasks are detected, the Invasion Detection System (IDS, p. 1 above) is of critical importance. However, current IDS architecture is usually cloud server-based, poses client data to significant privacy disruption and expands the time needed to detect potential intrusions. Deploying an advanced performance IDS implementing AI innovations on one unit turns out to be especially demanding when managing a huge amount of disseminated data across the infrastructure. In order to tackle these challenges, the above authors describe their new framework. A "fog-edge-facilitated Support Vector Machine (SVM)-federated learning for smart grid(s)" (p. 1 above). This type of learning indicates an AI method for edge-device training. In this innovation,

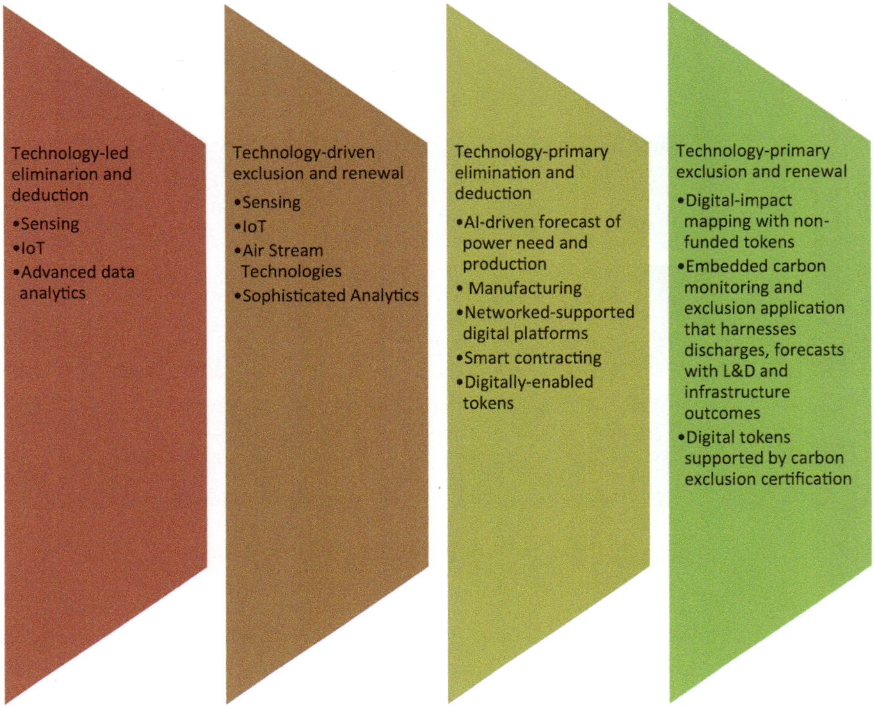

Fig. 1.7 Four technology-supported sustainability methods for net-zero implementation in energy domain

the learning boundaries are the only ones disseminated with the inclusive model, guaranteeing the highest privacy of data whilst permitting cooperative learning to create an IDS-quality based model. The prior authors in sharing with us the evaluation outcomes acquired from their suggested model, showcase its supremacy over current approaches, accomplishing a considerable advancement of "4.17% precision, 13.19% recall, 9.63% accuracy and 13.19% F1 score when assessed applying their selected dataset" (p. 1 above). In addition, their model achieved an excellent performance on their dataset, showcasing heightened accuracy, recall, precision and F1 metrics, obtaining 6.03%, 6.03%, 7.57% and 7.08% (as above), accordingly. Their innovative framework appears to advance the invasion recognition precision and ensures customer data privacy in smart grid(s), setting it as a considerable improvement in the domain.

The preceding authors continue to elaborate on their rationale by stressing that the smart grid(s) networks denote an improvement over traditional power (electricity) grid(s) with higher steadiness and capacity to provide companies and customers with undisrupted energy. These networks consist of an interaction and a power infrastructure between clients and energy businesses. The smart grid(s) network is built upon

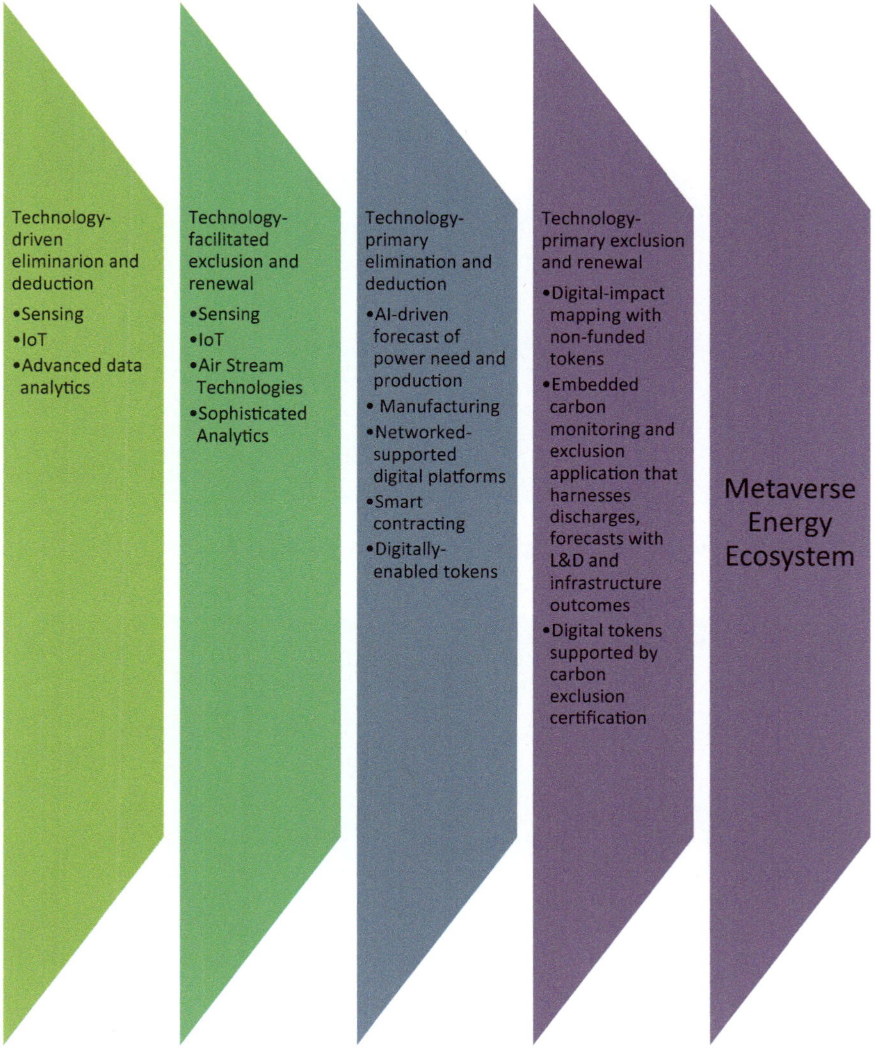

Fig. 1.8 The four technology-based sustainability methods for net-zero implementation envisioned for Metaverse Energy Ecosystem

the IMN innovation (as above), which includes smart metering and edge equipment, two-way interactions and a data ensemble cloud server to garner, handle data and implement monitoring assessments such as distance device screening in smart residencies (pp. 1–3 above). However, as the size of the conveyed data increases, the issues with the interaction infrastructure are also raised (p. 1 above). Appreciatively, 5G wireless interaction innovations progressively advance and provide with quick diffusion pace, an expanded frequency range interaction infrastructure and less

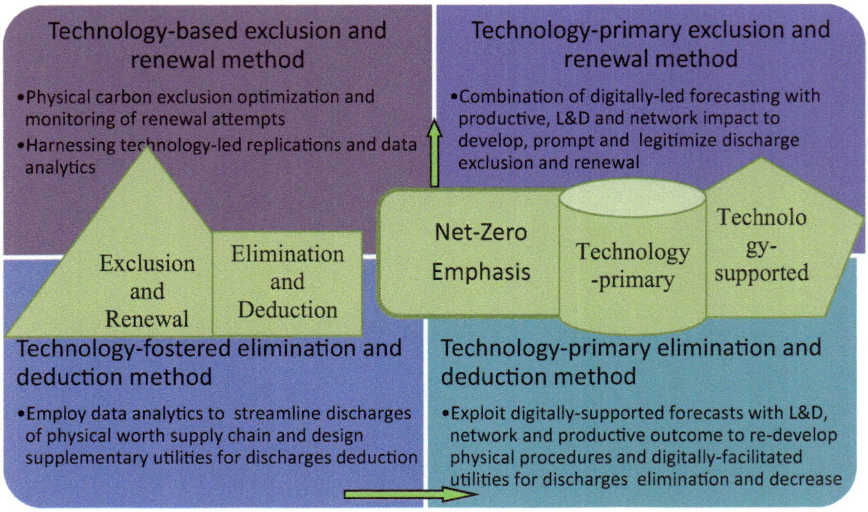

Fig. 1.9 A conceptual model outlining technology-led net-zero methods in energy sector

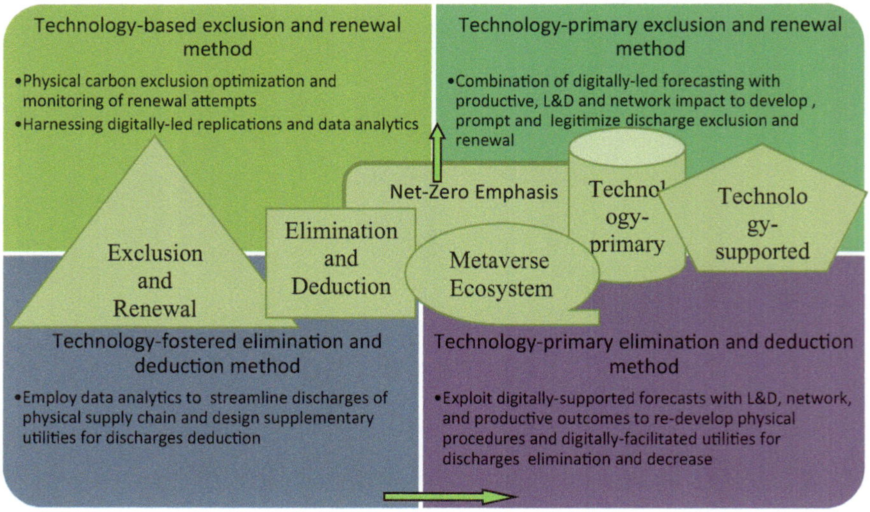

Fig. 1.10 The conceptual model of technology-based net-zero methods replicated on Metaverse energy space

broadcast wait. It appears, therefore, to set smart grid(s) and 5G integration as an extension pathway in the near future (pp. 1–2 above). Although the appearance of edge computing has been explored in some settings, Molokomme et al. [40] argue that the same insights cannot be rather claimed for the interconnections between edge intelligent innovation(s) and fog-computing on smart grid(s) in general and in Metaverse, in particular. Returning back to smart grid(s) ecosystem reflection on safety challenges, the first authors argue that implementing smart metering devices has developed adjustable tools and branched out IMN support by embedding wireless innovations into the smart grid(s). However, the IMN is susceptible to invasions due to expanded allocation, unsteady condition and two-way interaction infrastructure. The privacy and accessibility of the IMN structure might be debilitated by intrusions like "Denial of Service (DoS), Remote-to-Local (RtL), User-to-Root (UtR) and (or) Robing Attacks (RA)" [39, p. 20], inter alia. The aforementioned authors further emphasize the interaction challenges that relate to the central dolling out of the vast size of data produced by interrelated equipment, including necessities for data transmit, battery life-cycle, memory use and postponement. The smart grid(s) infrastructure, which is transitory, seems to be more sensitive to security dangers due to their dispersed modelling. Their decentralized status demands a rather scattered security system that serves scalability, interoperation, and adaptableness with a homogeneous security structure for diverse linked equipment.

Thus, fog-edge technology seems to provide dispersed support for delegating calculations of circulated smart grid(s) modelling. Federated learning (FL) [39, p. 5] offers its support to resolve the challenge of data confidentiality. Several clients train a joint model at the same time by giving out only the boundaries to cloud servers in federated learning as a replacement for untreated data. The extended usage of such safe and diffused modelling prompts for a privacy-maintained IDS for IMN. However, if the amount of customers grows, this element could trigger an important blockage as all distinct models are uploaded to the cloud concurrently, a condition that might disturb infrastructure functioning and grow resource usage.

In that respect, therefore, the preceding authors [39, 40] report on an intrusion detection system built upon the federated learning with a support vector machine and a level-based model with a fog cover. They prefer the support vector machine because of its recognized ability to deal with advanced-feature areas and sophisticated unsystematic resolution perimeters. Their reasoning for this proposal is to provide with a cutting-edge innovation that effectively recognizes invasions and addresses the growing concerns for data confidentiality in cloud server built-in invasion detection systems. Additionally, the federated learning permits diverse clients to jointly prepare a collective model without conveying unprocessed data. This safe and dispersed modelling lines up in a seamless way to ensure a privacy-maintained IDS for IMN, guaranteeing that customer data residue is safe and confidential. The authors stress the fact that their suggested method or modelling grants advanced invasion recognition operation, confidentiality maintenance, synchronized reaction and scalability in smart grid(s). It seems to fuel a new potential for safe and effective IDSs in the field of edge computation and IoT, where data confidentiality, suspension and asset barriers are core insights. In an overview of their proposed work, the prior authors emphasize the remote support vector machine cooperative modelling suggested, implementing FL whilst maintaining data confidentiality and addressing enhanced-featured space and unsystematic resolution restrictions. Also, the dispersed level modelling recommended employs a fog-edge level, disparate from a traditional cloud, for effective data dispensation, model preparation and decreasing suspension.

This level-based method might increase the scalability and reaction of the IDS, especially in vast-range and asset-restricted settings. In addition, the algebraic and proper illustration of diverse risk models, evidence of axiomatic proposals and client cases indicate the traits and behaviours of various invasion assaults for comprehending and alleviating safety risks in smart grid(s) infrastructure. Finally, the aforementioned authors focus on their scaled and comparison-based assessment illustrated, applying recognized indicators like accuracy, precision, recall, F1 score [39, p. 1] and preciseness between regional and universal and universal and up-to-date models.

The Smart Grid(s)

The smart or "intelligent" grid [39, p. 3] is considered to be a noteworthy digital-first innovation within the power (energy) network. The core aim of the traditional power grid(s) spans across effective electricity diffusion from central energy generation to extensive client basis. In specific, the notion of the smart grid(s) entails a pattern transfer from the conventional energy grid(s), which are based on an electromechanics monitoring structure, to a digitally first innovation of autonomous capacity infrastructure with remote screening abilities. This sophisticated grid architecture embeds diverse solutions, interaction platforms and monitoring operations to improve electricity production, diffusion, and expenditure. The smart grid(s) includes six core units: use supplier, diffusion, marketplace, client, function, and delivery.

The prior authors indicate that following recent research, the smart grid(s) consists of diverse digital sensing and ICT innovations, monitoring technology, information systems, and equipment that control multiple electrical tasks. The conventional grid model has received an important conversion owed to the amalgamation of smart grid(s) innovations and the appearance of functional issues. This shift is apparent across three critical levels, as next. First and second, advancing the diffused related data backside to administration and monitoring centres through an efficient response system, and third, often reacting in an autonomous way to regulate the response system's reaction appropriately. This development in grid(s) modelling has empowered its enhanced effectiveness and improved function. In addition, the preceding authors indicate that it has supported the application of more fine-tuned monitoring methods for successfully handling electricity production, diffusion, and expenditure. Also, it has fostered the management of data compilation and shift among diverse domain equipment and networks. Lastly, it has contributed to the analysis, assessment and support for gaining access to and exploiting data originated by digital-first innovations across the electricity grid(s). Charge forecast and matching, grid(s) trustworthiness evaluation, error recognition and control, and grid(s) safety against virtual attacks are interrelated challenges in smart grid(s).

The infrastructure modelling of the smart grid(s) entails three unified fields [39, 40], as next. First, the house area infrastructure (HAI). Second, the vicinity area infrastructure (VAI), and third, the broad area infrastructure (BAI). Each field assumes a clear responsibility in ensuring the seamless operation of the smart grid(s). Within this convoluted approach, tools like managerial monitoring and data attainment are placed within the extensive range of the BAI, guaranteeing effective control of the grid(s) on a generic range.

On the other side, the HAI and VAI possess the fundamental apparatus of the metering network, consisting of intelligent metering devices and data aggregators. These regional infrastructures prompt residencies and communities to take part in the underlying intelligent nature of the grid(s). Through the balanced interaction of these interrelated infrastructures, the smart grid(s) leads the way for a more digital-first, sustainable innovation energy network. Building upon their internal and external or inflow and outflow interaction interdependencies, we might move forward suggesting their open innovation (i.e., systems dynamic modelling) nature. In addition, the federated learning component strengthens policymakers to leverage concurrent outcomes, fostering timely and knowledgeable decisions. Further, it holds a considerable agency for shared intelligence by harvesting remote equipment and (or) information processing systems, turning it into a core element of this inspiring method. Embedding AI approaches has been revealed as a critical option with important advantages in smart grid(s) applications. Contemporary electrical grid(s) integrate dispersed elements of the smart grid(s) infrastructure in an undisrupted way, like sophisticated interaction modes, metering technology, and circulated energy (power) resources. These features survive with expanded energy infrastructure and complex interaction system(s), producing vast data. In such sophisticated solutions, the deliberate usage of AI strategies seems to become essential, supporting the efficient management, analysis, and extraction of meaningful outcomes from this data-empowered context. By leveraging the robustness of AI, the smart grid(s) tend to be in the position to accomplish optimized running, enhanced effectiveness, and regenerative improvements in energy, supporting power governance and transfer.

The Fog-Computing Layer in the Smart Grid(s)

The fog-computing layer seems to be a core feature of smart grid(s) technology. It appears to be embedded into the fundamental paradigm of smart grid(s) in an undisrupted way, offering a systems design and accommodating an effective space that addresses the sophisticated necessities of today's energy (power) networks. The primary principle of fog-computing level includes decentralized computational activities conventionally addressed by cloud computing, transferring them next to the data resources. This type of modelling, which focuses on edge-computing paradigm, increases data dispensation effectiveness and decreases suspension-based issues essential to guarantee the easy function of smart grid(s) [39, 40]. The fog-computing innovation originates from the cloud-computing technology that brings data and dealing out nearer to clients. It transfers computational tasks and data stockroom closer to the equipment itself rather than resting only upon central cloud-servers infrastructure. This approach moderates two core challenges with traditional cloud-computing innovation, as next. First, potential interruptions. Second, the prerequisite to forward vast amounts of data across extended continuum. It constitutes processing context-based, reflects actual solutions and support, feasible even when infrastructure connections are disrupted. This challenge tends to become more critical with the abundance of web-based equipment and the subsequent increase in the amount of data generated. In a nutshell, the fog-computing layer is considered to support businesses in their enterprise performance through edge-computing innovations and data

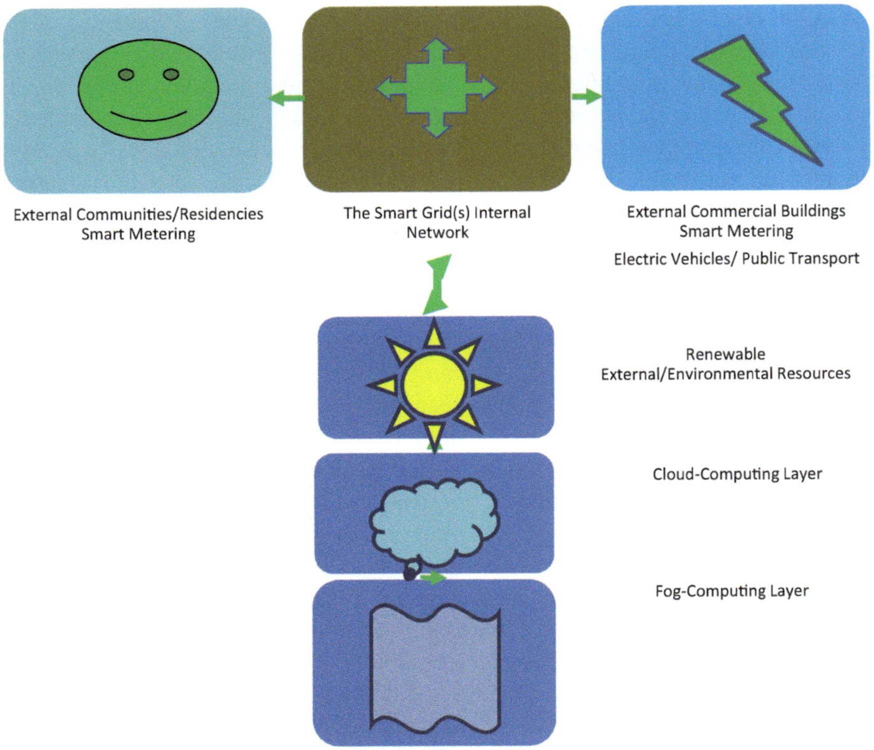

Fig. 1.11 The smart grid(s)-linked fog-computing layer on energy Metaverse

analytics. Extending the aforementioned authors' smart grid(s)-based fog-computing layer operationalization, Fig. 1.11 sketches an outline of an envisioned, "alternative future" fog-cloud smart grid(s) modelling on energy Metaverse that encompasses the notion of open innovation (external to internal environment-based agency).

The fog at the infrastructure's edge seems to enable remote data dispensation, decreasing suspension and increasing timely reaction. In the meantime, the cloud-computing paradigm tends to increase functioning by providing scalable stockroom and sophisticated analytics ability, contributing to data processing optimization, safety, and trustworthiness. The above authors continue their rationale by arguing that the fog-computing layer seems to become more fundamental in improving the advancement of the smart grid(s). Such grid(s) rely on timely and concurrent data processing from diverse resources, like sensing devices, smart metering, and climate conditions predictions. The smart grid(s) are able to increase energy (power) usage and decrease energy shortage by evaluating this data at the infrastructure's edge. Swift response time to supply chain and needs variations is fundamental to achieve these aims. Further, the smart grid(s) rest greatly on fog-computing links to maintain individual data safely. In that respect, the fog-computing level supports data safety by reserving data closer to its resource and restricting the requirement for a longer space transfer of data. This capacity seems to be a fundamental one in the smart grid(s) implementation, where data safety is of utmost significance. In addition, the prior authors claim that the fog-computing layer tends to be important for implementing the objective of the smart grid(s) for power (energy) effectiveness and reliability.

Due to concurrent and timely data dispensation and safe edge-computing level(s), the next-generation power (energy) infrastructure might be more reliable and safe with the assistance of the smart grid(s). The fog-computing layer reflects an intermediary one between the equipment ends and the cloud-computing one. It provides storeroom and timely dispensation to address the vast size of data generated by the smart grid(s) apparatus. It tends to foster swift processing and reduces suspension by acting as a mediator between the devices and decentralized cloud servers. Embedding a fog-computing layer into the smart grid(s) essentially advances functioning and trustworthiness by facilitating the concurrent dispensation and stockroom through decreasing suspension and lowering the gap between the district device(s) and the cloud-computing layer.

Recommended Architecture

The proposed modelling aims to advance the safety of the smart grid(s) against invasion attacks by using a remote invasion detection system (IDS). The prior authors outline their case study employed to support clarifying the suggested architecture which connects the fog-computing layer to federated learning to recognize invasions effectively and privacy-protecting [39, 40]. The model aims to accomplish scalability, concurrent, and timely recognition and to challenge invasion assaults by dispersing the invasion detection system's processing across smart metering devices and fog-computing layer. The architecture fosters data collection from various smart metering equipment whilst keeping data confidentiality due to the "engaged" (cooperative) status of federated learning. The universal fog-computing layer tends to act as a coordinator and aggregator agent rather than as a mere monitoring agent. The remote facet is formulated by administering neighbourhood models and training procedures across the improved metering network and fog-computing layer. The preceding authors [39, pp. 16–18, 40] elaborate on the core traits of their suggested architecture, as next. First, the allocation of neighbourhood models. This aspect is served in the remote invasion detection system modelling, where each fog-computing point supports a universal model that executes invasions recognition on the criteria (indicators, boundaries) garnered from the smart metering (processing neighbourhood models) within its region. These neighbourhood models are trained employing the support vector machine-linked federated learning, permitting the models to be trained from the data without requiring central data collection. The neighbourhood models are trained collectively, with up to date developments established through the universal server(s). Second, the facet of federated learning and cooperation. The first ensures the shared training of neighbourhood models without risking data confidentiality. Rather than transferring unprocessed data to the fog-computing layer, the neighbourhood models forward revised information or specifications to the universal server, collecting and harvesting the up-to-date information from all endpoints. This method seems to guarantee that confidential data is kept on the neighbourhood edges, maintaining confidentiality and decreasing the challenges related to central data stockroom. Third, the management and collection. This feature is supported through the training and shared (collaborative) model of up-to-date information from the neighbourhood model(s) (i.e., the enhanced metering network(s) performed by the universal server (i.e., the fog-computing layer). It garners the update employing confidentiality-preserving strategies, like safe collection or distinguishing confidentiality, to guarantee that each edge's contribution remains personal. Next, it is combined with the neighbourhood models so that they are able to be trained from their shared activity while reserving their traits. Fourth, the remote configuration. This modelling of the invasion detection system bears diverse wins. It splits the dispensation overload among various fog edges, increasing scalability and decreasing reliance on centralized server processing. It advances data confidentiality by monitoring vulnerable data and restricting information diffusion. Moreover, the cooperative status of the federated learning fosters comprehensive communication and architecture development among the neighbourhood models, ending up in increased recognition precision and adaptableness.

The above authors [39, 40] continue their argumentation by summarizing their case study on a remote invasion detection system for smart grid(s) performed. They emphasize the advancement of the safety of smart grid(s) against invasion assaults by applying a remote invasion detection system. Conventional invasion detection applications bear issues in scalability, confidentiality, and concurrent and timely investigation. Thus, a remote method employing fog-computing layer and federated learning is suggested by the authors to tackle these restrictions [29, 30, 39].

The Invasion Detection System Modelling

Analysing the above authors' remote invasion detection system modelling, it includes two core elements [39, p. 23], as next. First, the smart metering equipment placed at the client location that garners energy (power) expenditure information (data) and convey it to the fog level. Each smart metering device operates a neighbourhood invasion detection system to recognize discrepancies and possible invasions. Second, the fog level. It is a transitional computing layer between the smart metering and the centralized monitoring system. It consists of fog edges, which are regionally dispersed and proximate to the smart metering devices. Each fog edge operates in a neighbourhood invasion detection system architecture and manages the training with other edges in a federated learning approach. Its responsibility is to accumulate and assess the data garnered from the smart grid(s). It sustains a universal invasion detection system architecture that merges the awareness obtained from the neighbourhood models to decide upon invasion recognition accurately.

The Remote Invasion Detection System

This system consists of the following sub-parts, as follows. First, the neighbourhood architecture training. Each smart metering device processes regionally garnered information to train its neighbourhood invasion detection system model. This training stage encompasses support vector machine algorithms [12, 39] appropriate for managing advanced multifaceted component locations and irregular designs in invasion recognition. Second, the federated learning. The fog edges cooperate in a federated learning approach to develop the invasion detection system architecture, jointly. No more than the architecture indicators are transmitted as an alternative to exchange unprocessed information to maintain data confidentiality. Each fog edge corroborates to its neighbourhood architecture revisions to the collector model, where the universal invasion detection system architecture is upgraded, employing the federated learning aggregating. Third, the architecture allocation. The revised invasion and detection system model is separated back to the neighbourhood edges, enabling the recent knowledge being circulated across the remote invasion detection system. They then revise their neighbourhood models with the updated universal model, permitting them to adjust to developing invasion schemes. Forth, the concurrent and timely invasion recognition. Each smart metering device uses its revised neighbourhood invasion recognition model for synchronized invasion recognition. Inconsistencies, discrepancies, and possible invasions are recognized by matching up the detected metering information with the exercised schemes in the neighbourhood architecture.

Wins of the Remote Invasion Detection System.

The above authors [39, 40] argue that the use of a remote invasion detection system employing a fog-based federated learning model [39, p. 6] seems to present multiple wins [39, pp. 5–6], as next. First, the scalability. The remote modelling permits undisrupted scalability as newer smart metering devices can connect to the system without needing expanded adjustments. The federated learning framework guarantees that the invasion detection system is able to be supported from the shared awareness of a vast amount of contributing smart metering devices. Second, the confidentiality maintenance. By harvesting the federated learning, the invasion detection system keeps data confidentiality. The unprocessed metering information is maintained regionally and only cumulative model revisions are divided across the infrastructure. It cares for vulnerable client data whilst permitting the system to be trained from multiple data resources. Third, the synchronized recognition. With the invasion detection system architecture exercised at the improved metering network and fog level, concurrent invasion recognition can be accomplished. The immediacy of fog edges to smart metering devices permits quicker assessment and reaction, decreasing the recognition suspension in relation to a central framework. Fourth, the durability. The remote invasion recognition system advances hardiness by preventing error on its own. Even if some fog edges or smart metering devices turn into settlement, the shared intelligence of the dispersed invasion recognition models are able to lend support to efficient recognition and alleviation.

The Federated Learning-Led Invasion Detection System

In this part, the prior authors [39, pp. 2–4, 40] collectively outline the suggested strategy for the recognition of invasion for smart grid(s) systems. The core traits of this system are envisioned as follows. First, each customer's neighbourhood support vector machine. Each user in the federated learning network assumes its own neighbourhood support vector model. The neighbourhood support vector model receives training by means of the customer's neighbourhood data, which includes information not known to other customers or the main server. The neighbourhood support vector model is trained on the schemes and traits of the information particular to each customer, permitting it to make forecasts regionally. Second, the neighbourhood model indicators (or criteria). Following the neighbourhood support vector model training, the customer acquires the neighbourhood model criteria (boundaries) related to its support vector model. These indicators depict the well-read loads and preconceptions of the support vector model, which identify the decision edge and facilitate it to categorize newer information. Third, the interaction with the server. The neighbourhood model criteria are forwarded from each customer to the core server. This interaction can be performed safely, enabling the confidentiality of the neighbourhood information and architecture. Fourth, the universal support vector machine and the universal model criteria. Hovering over the server, the obtained neighbourhood architecture boundaries are cumulated and employed to revise the universal support vector machine. They reflect the shared awareness of all customers and intend to hold a broader depiction of the information. The server calculates the universal model criteria (universal model parameters) rested upon the cumulated neighbourhood model boundaries. Fifth, the architecture criterion interaction and iteration-based learning. The revised universal model criteria are sent back to the customers for additional learning (training) and architecture development. This criterion interaction and iteration-triggered learning procedure permits the customers to gain from the shared awareness of the whole ecosystem whilst maintaining the confidentiality of their neighbourhood information. The iteration-led learning resumes in anticipation of the universal support vector machine reaching a desirable range of precision or a particular amount of repetitions achieved. By harvesting this federated learning model, many customers are able to cooperatively train a support vector architecture whilst maintaining their neighbourhood information confidential. The centralized server supports the collection of neighbourhood architecture criteria, permitting awareness distribution without immediate information interaction. This framework fosters confidentiality, data safety, and dispersed training in cases when information is not possible to be formally kept or joined due to confidentiality challenges or law regulations. The assessment and validation of the end result federated learning architecture seem to be critical to evaluate its outcomes and define its efficiency in recognizing invasion. As the authors report above, a detached assessment data compilation is employed to assess the architecture as assessment information for both compilations. The outcomes are evaluated according to varied indicators, like precision, accuracy, recall or F1 scores (elaborated subsequently). When trained, the architecture can be positioned and utilized to recognize invasions on newer information without additional tagging. This is due to the fact that the architecture has been trained to recognize the information schemes connected to invasions.

Evidence Assessment

The simulated outcomes are explored to obtain knowledge about the operation of the suggested model in invasion recognition using corresponding datasets. The acquired outcome indicators are weighed against point of reference evidence or obtainable invasion recognition methods to assess the efficiency of the recommended framework. The aforementioned authors [39, 40] elaborate on their outcomes and rationale, by stressing that their suggested methods tend to be depicted by their remote and dispersed approach, data-confidentiality keeping and integration of a fog-computing layer. Their method showcases considerable capacity in managing distinctive safety needs whilst lessening confidentiality and suspension challenges in smart grid(s). Further exploration and development in this domain might end up in the design of tangible, durable, and effective invasion detection models for the smart grid(s) infrastructure. One of the core wins of this approach is its dispersed and remote modelling. Against various central invasion recognition systems that rely on a mere level of monitoring or data storage, this framework uses federated learning to permit many sites to train and advance the invasion recognition architecture cooperatively. This method advances the scalability of the infrastructure. It reduces confidentiality challenges with central patterns, in which a centralized server might be admitted to use vulnerable information and greatly susceptible to mere error. The embeddedness of the fog-computing layer improves the effectiveness and response of smart grid(s). The fog-computing layer fosters the performance of data processing and decision latitude activities at a node of an ecosystem, ending up in lowered suspension in spotting safety risks in nearby immediacy to clients. Prompt invasion recognition and reaction in smart grid(s) is critical since timed and concurrent management can diminish possibly damaging web-assaults.

Moreover, the federated learning ensures that vulnerable client information is maintained safe on the neighbourhood architecture (or server). The latter is able to be trained without offering information to the centralized server, constituting it as an advanced alternative for confidentiality-tuned smart grid(s) settings. Thus, it lines up with the developing importance of preserving client data in smart grid(s). Whilst these wins tend to constitute the suggested support vector machine-linked federated learning with the fog-computing layer an undeniable option for smart grid(s) invasion recognition, it is necessary to recognize possible issues and restrictions. These might entail challenges connected to confluence, interaction upwards, and the necessity to achieve a cautious architecture upgrade. Whilst the fog-computing layer provides with the win of neighbourhood data dispensation and faster decisions taken at the infrastructure node, it might well present system interruptions in particular cases. The fog-edges might have diverse operational weight than central servers, pending on the convolution of the invasion recognition algorithms and the size of information being handled, where there could be an exchange between suspension and computational interval. Matching the computational challenges with decreased reaction tends to be a core issue in employing the fog-computing layer for the invasion recognition in smart grid(s). Moreover, the federated learning rests on interaction between fog-computing edge equipment and a centralized server, where the architecture modifications are collected and distributed. This interaction might host higher ground in relation to infrastructure volume and power (energy) expenditure, especially in conditions with recurrent architecture alterations. Addressing this interaction projecting effectively whilst enabling architecture confluence, seems to be not an easy activity. The federated learning usually entails machine learning training across dispersed and possibly mixed information resources. Making sure that the models join a coherent and precise portrayal of the invasion recognition activity, might well be demanding. Inconsistency in information transmission, excellence or equipment capacity across smart grid(s) edges might prevent architecture junction. Cautious improvement and management need to be performed to handle this restriction. As the smart grid(s) infrastructure extends, the scalability of the suggested method appears as an issue. Arranging for and sustaining the essential properties, including the fog-computing edges and equipment, could be demanding. Making sure that the network balances efficiently with the development of the smart grid(s) system requests meticulous preparation and running. Likewise, not all edge equipment in smart grid(s) might have the capacity to take part in the federated learning successfully. Guaranteeing congruence and optimization model for resource-restricted equipment appears to be critical in accomplishing expanded usage of the recommended method.

Whilst the suggested strategy is targeting at advancing safety, it further reveals newer related concerns. Defending the trustworthiness of architecture modifications during interaction and protecting possible counter-balanced assaults on the federated learning procedure, are essential issues that need to be managed. In addition, the smart grid(s) data might showcase discrepancies and notion flow, where usual and invasion information dispersion and traits alter in due course. The method should be tuned to adjust to these modifications efficiently and prevent from being outdated. Whilst the suggested support vector machine-linked federated learning with the fog-computing layers provides essential wins for invasion recognition in smart grid(s), it seems to be necessary to recognize and handle the aforementioned restrictions. Real-time testing experimentation tends to be vital for authorizing the efficiency of the architecture in existing smart grid(s) contexts, notwithstanding the insightful updates offered by the corresponding replications. Moreover, the efficiency of the prospective invasion recognition system might be optimized by exploring the complementary machine learning architecture. Furthermore, conquering these issues is expected to demand combined computational developments, effective interaction modus operandi, resource management, and continuous exploration to guarantee the method's efficiency and scalability in real-time smart grid(s) settings. These features tend to stress the demand for additional exploration and evidence-based verification to empower the flexibility and durability of the prior authors' recommended method. Decreased recognition precision, increased error alarm, and restricted marked data accessibility test the new invasion recognition methods. Provided the rising complexity of web assaults, empowered invasion recognition systems exercise an essential agency in protecting the infrastructure and vulnerable data. The authors conclude that their exploration expands its emphasis on the sphere of smart grid(s), where the interrelation of equipment and information exchange sets important safety challenges. Harvesting the federated learning and the fog-computing layer, their suggested approach provides a remote framework for defining invasions within the smart grid(s) contexts. Their framework optimizes current techniques through expanded assessment, standing out in precision, recall, accuracy, F1 scores, and explicitness for recognizing and categorizing infrastructure invasions. It showcases adaptableness to optimizing assault designs, scalability across differentiated system contexts, and an efficient equilibrium between accuracy and recall. In their future research, they intend to investigate various infrastructure conditions, developing assault cases and differentiated information resources, while advancing real-time agency and embeddedness with existing safety approaches in a seamless way. This tends to spot an essential level towards empowering the safety of smart grid(s) in the event of developing web-assaults. Further to the aforementioned plans, the authors are expected to incorporate the B technology in the suggested ecosystem for advanced scalability and safety. They plan to expand it to fundamental network and the IoT technology to create sheltered, differentiated, and interrelated networks.

1.7 Conclusion

The Metaverse-led future open research challenges seem to be rather vast and multimodal in nature. Related evidence calls for greater attention for, exploration of, co-design and assessment of immersive innovations and smart intelligence parameters that uphold the smart grid(s) on power Metaverse open innovation infrastructure through renewable resources and sustainable governance. A liveable smart grid(s) on energy Metaverse might be sustainable through renewable resource-led power generation, storage, diffusion and expenditure, backed up by sensing and fog-computing technologies that might address safety, management and control issues successfully.

References

1. Ojo O, Dada T, Ademola Y, Olurankinse G, Ebuka E, Acka B, Kujah F (2025) Smart grids and IoT-enabled renewable energy integration. Path Sci 11(1):8012–8022. https://doi.org/10.22178/pos.113-28
2. Stanchev P, Hinov N (2026) Smart grids and sustainability in the age of PMSG-dominated renewable energy generation. Energies 19:772. https://doi.org/10.3390/en19030772
3. Al Khafaf N, Song H, Kamoona A, Sabar N, McGrath B, Yu X, Jalili M (2026) Smart meter data intelligence for sustainable distribution network operations: state-of-the-art applications and pathways toward net-zero. Renew Sust Energ Rev 231:116723. ISSN 1364-0321. https://doi.org/10.1016/j.rser.2026.116723
4. Martínez-Gutiérrez A, Díez-González J, Perez H, Araújo M (2024) Towards industry 5.0 through metaverse. Robot Comput Integr Manuf 89:102764. https://doi.org/10.1016/j.rcim.2024.102764
5. Salloum S, Al Marzouqi A, Alderbashi KY, Shwedeh F, Aburayya A, Al Saidat MR, Al-Maroof RS (2023) Sustainability model for the continuous intention to use Metaverse technology in higher education: a case study from Oman. Sustainability 15:1–19. https://doi.org/10.3390/su15065257
6. El Jaouhari A, Arif J, Samadhtiya A, Anil Kumar A, Trinkūnas V (2023) Are we there or do we have more to do? Metaverse in facility management and future prospects. Int J Strateg Prop Manag 27(3):159–175. https://doi.org/10.3846/ijspm.2023.19516
7. Bigliardi B, Ferraro G, Filippelli S, Galati F (2021) The past, present and future of open innovation. Eur J Innov Manag 24(4):1130–1161. https://doi.org/10.1108/EJIM-10-2019-0296

8. Wyrwich M, Steinberg PJ, Noseleit F, de Faria P (2022) Is open innovation imprinted on new ventures? The cooperation-inhibiting legacy of authoritarian regimes. Res Policy 51(1):104409. ISSN 0048-7333. https://doi.org/10.1016/j.respol.2021.104409
9. Ramírez-Herrero V, Ortiz-de-Urbina-Criado M, Jose-Amelio Medina-Merodio, JA (2024) Understanding the knowledge structure and the value creation process of the metaverse. Heliyon 10(10):e31271. ISSN 2405-8440. https://doi.org/10.1016/j.heliyon.2024.e31271
10. Soliman MM, Ahmed E, Darwish A et al (2024) Artificial intelligence powered Metaverse: analysis, challenges and future perspectives. Artif Intell Rev 57:36. https://doi.org/10.1007/s10462-023-10641-x
11. Ghazinoory S, Parvin F, Saghafi F, Afshari-Mofrad M, Ghazavi N, Fatemi M (2025) Metaverse technology tree: a holistic view. Front Artif Intell 8:1545144. https://doi.org/10.3389/frai.2025.1545144
12. Shen J, Zhou X, Wu W, Wang L, Chen Z (2023) Worldwide overview and country differences in Metaverse research: a bibliometric analysis. Sustainability 15(4):3541. https://doi.org/10.3390/su15043541
13. Chen Z, Gan W, Sun J, Wu JS, Yu PS (2024) Open Metaverse: issues, evolution, and future. In: Companion proceedings of the ACM web conference 2024 (WWW'24). Association for Computing Machinery, New York, NY, pp 1351–1360. https://doi.org/10.1145/3589335.3651898
14. Iqbal MZ, Campbell AG (2023) Metaverse as tech for good: current Progress and emerging opportunities. Virtual Worlds 2(4):326–342. https://doi.org/10.3390/virtualworlds2040019
15. Lowry PB, Boh WF, Petter S, Leimeister JM (2025) Long live the Metaverse: identifying the potential for market disruption and future research. J Manag Inf Syst 42(1):3–38. https://doi.org/10.1080/07421222.2025.2455770
16. Karaarslan E, Aydın Ö, Yazıcı Yılmaz S, Tınmaz H, Eken S, Gökçe Narin N, Cali U, Özkaynar K, Gürcan Ö, Fanea-ıvanovici M, Güçlütürk OG, Qi M, Kusetogullari H, Challenger M, Bacanin N (2025) Shaping the future of Metaverse research: innovation, collaboration, and journal of Metaverse's academic impact. J Metaverse 5(2):73–91. https://doi.org/10.57019/jmv.1664054
17. Karunarathna S, Wijethilaka P, Ranaweera KT, Hemachandra TS, Liyanage M (2023) The role of network slicing and edge computing in the Metaverse realization. IEEE Access 11:25502–25530. https://doi.org/10.1109/ACCESS.2023.3255510
18. Patra A, Pandey A, Hassija V, Chamola V, Mishra RP (2024) A survey on edge enabled Metaverse: applications, technological innovations, and prospective trajectories within the industry. IEEE Access 12:125125–125144. https://doi.org/10.1109/ACCESS.2024.3452184
19. Alsharif MH, Kelechi AH, Jahid A, Kannadasan R, Singla MK, Gupta J, Geem, ZW (2024) A comprehensive survey of energy-efficient computing to enable sustainable massive IoT networks. Alex Eng J 91:12–29. ISSN 1110-0168. https://doi.org/10.1016/j.aej.2024.01.067
20. Yakubu MM, Hassan MFB, Danyaro KU, Junejo AZ, Siraj M, Yahaya S, Adamu S, Abdulsalam K (2024) A systematic literature review on Blockchain consensus mechanisms' security: applications and open challenges. Comput Syst Sci Eng 48(6):1437–1481. https://doi.org/10.32604/csse.2024.054556
21. Tightiz L, Dang M, Padmanaban SK, Hur K (2024) Metaverse-driven smart grid architecture. Energy Rep 12:2014–2025. ISSN 2352-4847. https://doi.org/10.1016/j.egyr.2024.08.027
22. Fan Y, Zhang L, Zhang C, An Z, Li B, Li D (2025) Fully printed and scalable current and voltage sensors for smart grid transmission line monitoring. Sensors 25(7):2287. https://doi.org/10.3390/s25072287
23. Tungturu SD, Karthik S, Boumhaout Y, Elfikky A (2025) The Metaverse: a comprehensive survey of technological foundations, societal implications, and future research directions. Preprints:101971. https://doi.org/10.20944/preprints202510.1971.v1
24. Zuo J (2024) Challenges of new energy power systems. Int J Electr Power Energy Stud 2(3):20–24. https://doi.org/10.62051/ijepes.v2n3.04
25. Girgibo N, Hiltunen E, Lü X, Mäkiranta, AVT (2024) Risks of climate change effects on renewable energy resources and the effects of their utilisation on the environment. Energy Rep, 11:1517–1534. ISSN 2352-4847. https://doi.org/10.1016/j.egyr.2024.01.024

26. Ahmad F, Boumaiza A, Sanfilippo A, Al-Fagih L (2025) A detailed comprehensive role of digital technologies in Green Finance Initiative for net-zero energy transition. Adv Energy Sustain Res 6:2500066. https://doi.org/10.1002/aesr.202500066
27. Agrawal S (2025) NetZero emissions: current trends, barriers, and opportunities. Univ Res Rep 12(2):142–148. https://doi.org/10.36676/urr.v12.i2.1535
28. Stern N, Valero A (2021) Innovation, growth and the transition to net-zero emissions. Res Policy 50(9):104293. https://doi.org/10.1016/j.respol.2021.104293
29. Cao L, Hu P, Li X et al (2023) Digital technologies for net-zero energy transition: a preliminary study. Carbon Neutral 2(7). https://doi.org/10.1007/s43979-023-00047-7
30. Falcke L, Zobel AK, Comello SD (2024) How firms realign to tackle the grand challenge of climate change: an innovation ecosystems perspective. J Prod Innov Manag 41(2):403–427. https://doi.org/10.1111/jpim.12687
31. Uchenna Attah R, Baalah Garba MP, Ogunsola OY (2023) Advances in sustainable business strategies: energy efficiency, digital innovation, and net-zero corporate transformation. Icon Res Eng J 6(7):450–469
32. Virmani N, Kumar P, Karuppiah K, Jagtap, S (2025) Adopting net-zero supply chains: integrated model of TOE and TAM in emerging economy context. Sustain Fut 10: 101188. ISSN 2666-1888. https://doi.org/10.1016/j.sftr.2025.101188
33. Liu B, Wang F (2025) The impact of the global digital economy on carbon emissions: a review. Sustainability 17:5044. https://doi.org/10.3390/su17115044
34. Hameed M, Abdulkareem Hameed N, Natiq Abdulwahab I, Hashim Qasim N, Saad SA (2025) Drones and IoT for enhancing renewable energy integration. Int J Inform Vis 9(5):2094–2103. https://joiv.org/index.php/joiv/article/view/4630/1461
35. Xie H, Liu Y (2025) The application of digital technologies in carbon emission reduction in China's energy and power industry. Net Zero 1(1):6–19
36. Liu Z, Huang Y, Mai J, Li W, Cai Z, Li Y (2025) Is the metaverse really coming to fruition? A survey of applied metaverse and extended reality. High-Confidence Comput 100376. ISSN 2667-2952. https://doi.org/10.1016/j.hcc.2025.100376
37. Li J, Shirowzhan S, Pignatta G, Sepasgozar SME (2024) Data-driven net-zero carbon monitoring: applications of geographic information systems, building information modelling, remote sensing, and artificial intelligence for sustainable and resilient cities. Sustainability 16:6285. https://doi.org/10.3390/su16156285
38. Vishwanadham Mandala V et al (2024) Towards carbon-free automotive futures: leveraging AI And ML for sustainable transformation. Educ Admin Theory Pract 30(5):3485–3497. https://doi.org/10.53555/kuey.v30i5.10500
39. Tariq N, Alsirhani A, Humayun M, Alserhani F, Shaheen M (2024) A fog-edge-enabled intrusion detection system for smart grids. J Cloud Comput 13(43):1–34. https://doi.org/10.1186/s13677-024-00609-9
40. Molokomme DN, Onumanyi AJ, Abu-Mahfouz AM (2022) Edge intelligence in smart grids: a survey on architectures, offloading models, cyber security measures, and challenges. J Sens Actuator Netw 11(47):1–50. https://doi.org/10.3390/jsan11030047
41. Huang J, Sun P, Zhang W (2022) Analysis of the future prospects for the Metaverse. In: Proceedings of the 2022 7th international conference on financial innovation and economic development (ICFIED 2022). Atlantis Press, pp 1899–1904. https://doi.org/10.2991/aebmr.k.220307.312

Chapter 2
Metaverse Intelligent Energy Sector Management-*SmartEnergyverse* Management

Abstract In the current worldwide smart innovations aligned with sustainable energy management, there tends to be extensive research, development, innovation, enterprise, impact, community engagement and stakeholder management. However, the above cutting-edge sustainable smart energy research and innovation investment is seldom explored or applied into the envisaged, future Metaverse ecosystem. This could bear feasible, actionable, accessible, scalable, computationally powerful, regulatory, decentralized, secure, accredited, sustainable-driven and energy-and-renewable energy efficient return on investment (ROI) for future Metaverse smart energy sector ("SmartEnergyverse") management, globally. Thus, the present chapter highlights this expanded gap in a multi-fold array, as follows. First, by exploring and reviewing the connected, yet rare, empirical-driven research, innovation, enterprise and impact (PIEI) across the smart, though energy-demanding intelligence models, technologies or solutions adhered to the "smarter" Metaverse energy domain. These smart energy innovations might embrace multimodal examples, starting from Digital Twin (DT), Internet of Things (IoT), Artificial Intelligence (AI), Virtual Reality (VR), Machine Learning (ML), Augmented Reality (AR), Extended Reality (XR), Computer Vision (CV) and extending to Smart Sensing (SS), Blockchain (B) and Ethereum (E), Proof-of-Work vs. Proof-of-Stake (PoW vs. PoS), Edge and Cloud Computing (ECD), Data Centers (DC) and Communication and Functional Optimization (CFO), among others. Second, by bringing to light the potential of these smart energy models or innovations to be optimized to serve an energy-and-renewable energy effective, envisioned Metaverse management infrastructure. Third, by highlighting the enablers (wins) and barriers (concerns) the above intelligent energy models or innovations might entail in order to support a global, envisioned "SmartEnergyverse" management ecosystem across multi-fusion aspects. For instance, beginning from sustainable research and development (SRD), to monetization, green knowledge performance indicators (GKPIs), user quality interaction (interface) and experience (UQI and UQX), scaling, safety, privacy, regulations, accreditation and extending to trade-offs between energy-led Metaverse system implementation and performance (e.g., smart energy and eco-friendly energy generation, maintenance, distribution, peer-to-peer consumption for blockchain energy marketplace, for suppliers, consumers, regulatory bodies, energy factories, energy

E. G. Makri, *Metaverse and Smart Grid in Energy Industry: An Open Innovation Challenge on Trends and Applications*, Studies in Smart Technologies,
https://doi.org/10.1007/978-3-032-29666-5_2

communities, etc.). Fourth, by operationalizing the smart energy intelligent models and innovations embedded into the "SmartEnergyverse Management" as "mirroring" an open and eco-innovation (i.e., systems design thinking, outflow to inflow exercise of innovation, co-design, environment-friendly, etc.) blockchain-led, local energy supply chain management ecosystem. Fifth, by outlining the need to develop smart intelligence energy-ROI-driven research, innovation, enterprise and outreach within an eco-friendly, envisaged Metaverse infrastructure.

Keywords Smart intelligence · "SmartEnergyverse" management · Open and eco-innovation · Renewable resources · Blockchain · Ethereum · Proof of stake · Proof of work

2.1 Introduction

Zhang et al. [1] indicate that the energy-demanding nature of the intelligent technologies associated with energy Metaverse [e.g. DT (development and operation), B (data repository and cross-over functionality), IoT (connecting physical objects to the Web), Internet of Energy (IoE) (linking people, procedures, data and objects) [2], AI-led (independence and smart intelligence) and (or) XR-driven (immersive UQI and UQX endorsing VR, AR and MR, etc.), seem to instigate reasonable concerns over its sustainable operation. In their present work, they suggest the so-called "greener Metaverse ecosystem" as means to optimize the energy effectiveness of the energy domain Metaverse, in a generic way, as follows. They first explore the energy-demanding intelligent systems on Metaverse. Then, they elaborate on the computational and networked aspects (e.g., semantics, stochastic computing) (p. 1 above) of the energy-value added advanced solutions embedded into the energy landscape Metaverse. Next, they present their use case of energy-effectiveness transportation on energy landscape Metaverse. Lastly, they highlight the core issues that the "Greener Metaverse" sustainable development might entail, in order to trigger additional RIEI across this eco-innovation, energy domain Metaverse. At the end, the authors outline a summary of the 4-layer Metaverse energy infrastructure and connected energy-"hungry" procedures. This synopsis of the 4-tier Metaverse energy ecosystem that connects energy-demanding processes is presented in Fig. 2.1, as an extended version that integrates our own tentative, open innovation-driven scaling operationalization (or reflection) suggested (i.e., outflow to inflow innovation synergy). The cross-over angles of the following flow chart are highlighted in further outlined details in the following sections that come next.

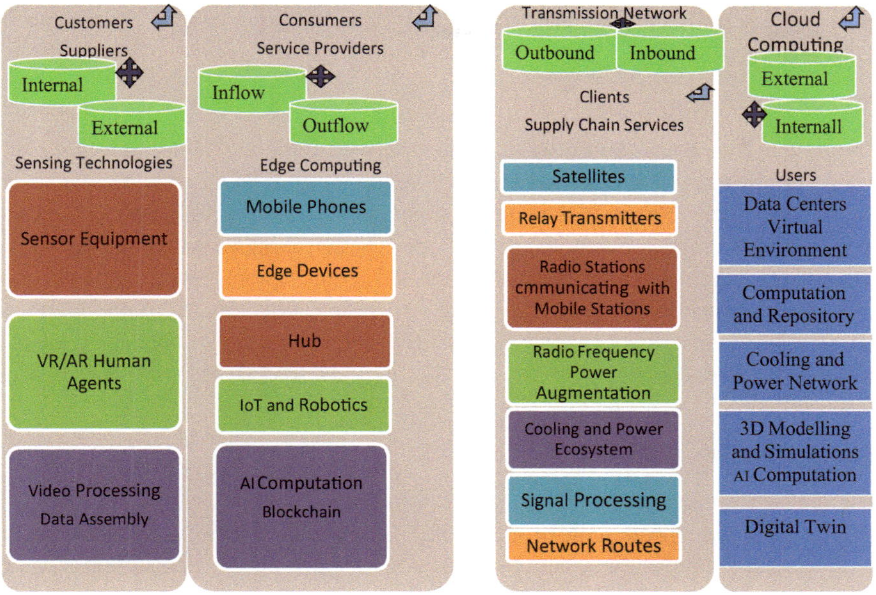

Fig. 2.1 A 4-layer open innovation-led energy domain Metaverse infrastructure linked to energy-intensive technologies

2.2 Metaverse, Advanced Computational Technologies and Energy Sector

Elaborating on their recent overview that bridges current scarce insights and evidence-based accounts on existed less greener, intelligent Metaverse energy sector properties, Rafique and Qadir [2] highlight the following enlightening review, as next.

Digital Twin—DT

The associated energy domain cloud-led solutions seem to deliver heightened computational power and energy-"hungry" functions [e.g., 3-dimensional (3D) modelling and simulations, AI-empowered pre-trained models, etc.] to high-performance application programming interfaces (APIs) in remote data centers when the ability of the local energy ecosystem is rather limited. However, the intensive dependence on cloud-based energy systems could aggravate the energy-driven computational overhead. For example, the energy-demanding data centers resting on their apparatus, decreasing heat and network equipment, commonly use 40–55% of the total energy utilization. In order to offset the excessive energy needs, extensive energy adjustments are made to the APIs and the energy infrastructure's foundation. Conducive to the above, indicative examples of present tech giants' data centers (e.g., Google) are perceived to sustain a power effectiveness of 1.1 at the time when the regular knowledge performance indicators (KPIs) for such ecosystems tend to be 1.57 (pp. 3, 6, 13 above).

Artificial Intelligence—AI

The AI-empowered systems provide Metaverse with autonomy and smart intelligent energy applications that tend to heighten the energy sector UQI and UQX spectrum. Despite the gigantic pace of the emerging generative AI (GenAI) solutions and the longer-term emphasis on how their system performance might be best optimized for energy usage, their challenging issues of how they could become less expensive to run and establish greater energy effectiveness across the energy context, seem to appear as rather non-consensus ones, yet. The latest trends to generate sustainable energy evaluation KPIs across ML models and AI solutions, tend to be rather restricted in range and depth.

Extended Reality—XR and Blockchain—B

The above authors argue that the energy effectiveness-KPIs which align with the UQX energy sector level, seem to trigger the design of VR applications with advanced visual and eco-innovation system performance (i.e., less energy demanding). The B platform(s) are considered as the gigantic storage providers of the enormous data generated on energy Metaverse. They run in a decentralized mode, are scalable, decrease the transfer of data and guarantee safety and transparency. The deployment of cross-over chains permits data transaction between Bs and optimizes a more interoperable energy domain Metaverse. Non-fungible tokens (NFTs) (p. 7 above) are perceived as exclusive cryptocurrency tokens on Bs that showcase the possession and trading records of digital or actual energy assets. This type of tokens can ascertain distinctiveness, monitor energy ownership-status, develop trusted energy interaction and improve transaction automation through energy smart contracting, as well. Cryptocurrency B energy assets, on the other hand, seem to restrict energy exchange fee(s), improve safety and confidentiality and shorten energy money transmission. However, the above authors argue that there tend to be still remaining system performance and energy expenditure challenges associated with such solutions that they believe they have not been adequately addressed, yet. The first of them reflects the programming and procedure followed to accomplish dispersed consensus about the connected ledger energy solutions or related data sets (i.e., "consensus mechanism or proof-of-work", p. 6 above). This condition runs on intense computational power architecture exercised on most cryptocurrency energy Bs with digital energy remains. The second one adheres to the case when duplicate pathways between energy nodes travelling through and the number of the energy nodes across the B energy ecosystem rising fast, end up in heightened energy "redundancy" (p. 7 above). To assess the total energy effectiveness, the energy expenditure per exchange is often applied to weigh energy Bs against other digital energy innovations. Resting on this KPI, however, might be rather confusing, as the connected energy consumption principally relies on the aforementioned energy consensus mechanisms and (or) proof-of-work (PoW) and not on the energy exchange capacity. The total volume of energy expenditure remains important for the present energy Bs, but it is expected to be modified once the

Ethereum cryptocurrency energy B moves from the PoW to the proof-of-stake (PoS) energy-efficient or green(er) energy Metaverse status. In sum, the above authors outline that the scalable-led energy sector Metaverse needs energy-effective digital innovations to remain sustainable. However, in certain cases, any trade-off demands between system performance and (or) computational power and scalability, need to be proportionate to energy-effectiveness Metaverse modelling and UQI and UQX energy landscape practice.

2.3 Emerging Intelligent Systems Reflecting Energy Effectiveness for Green(er) Energy Sector Metaverse

Digital Twin-Sensing and-Edge Ecosystems for Energy Efficiency

In IoT energy scene infrastructure, the garnered data are centralized and directed towards either cloud and (or) locally linked access route. Within this domain, the main category of energy decrease tools rests within sensor(s). Those inactive, they spend redundant power. Therefore, an effective energy expenditure set up is requested to lower energy waste, while fulfilling quality services. The prior authors note that 30–50% of the energy decrease is achieved by timetable (or scheduled) optimization and volume of work allocation, supported by sensing tools that promote the development of energy data transfer effectiveness. The IoT sensing devices might use cost-free energy options (e.g., radiance, radio wave, thermal amplitude or motion energy), with other energy-efficient schemes leveraging renewable resource management. Research-driven energy-effective algorithms are expected to spend 40–80% decreased power with optimized energy level and lower cost. As an alternative, edge computing equipment might leverage the energy offered by wireless power transfer networks, while an innovative-led algorithm was prorun to optimize the system performance and energy expenditure of the wireless powered ecosystem with a drastically limited processing timeline.

Communication Infrastructure—CI

The prior researchers further note that the ease of access and ability to move and expand (i.e., scalable, flexible) elements tend to constitute crucial properties for the energy field Metaverse. Relevant reviews have identified next generation "green radio" (pp. 2, 4, 16 above) ecosystems that can provide indicative examples of energy system optimized performance, as follows. For instance, diverse platforms connect different systems hierarchically to serve the matching users and smaller cells are aligned to advance network output and decrease energy expenditure. Or else, cell-zoom features a supporting traffic optimization and energy idle mode, electromagnetic radiation vast volume inflow-and-outflow developments that improve the system's performance and energy effectiveness and equipment connections without passing through the energy infrastructure and (or) 6G AI-fostered energy data transmission (p. 1 above), inter alia.

Data Centers and Cloud Computing—DC and CC

Along other lines, the same authors claim that an inactive server tends to consume 2/3 of energy when entirely used. Within Metaverse, a physical device is able to host many virtual energy ones (i.e., energy servers) separately to prevent suboptimal usage that increases energy effectiveness, in parallel. This strategy of merging insufficient usage energy servers represents the so-called "server consolidation" (p. 11 above) that is supported by energy effectiveness configuration. Optimizing the physical and virtual energy mapping formulation tends to denote the virtual-Metaverse status across the energy territory, as a challenge. Facilitating the shared optimization of the latter with the corresponding energy task schedule-based utilization might serve a more energy-efficient ecosystem that can be further assisted by more energy-friendly AI-empowered cooling infrastructure (e.g., Google reducing 40% of their energy expenditure).

Machine Learning—ML

An effective ML architecture adoption that rests on energy modelling and data compression and refining (i.e., decreasing the size of a large model without training a newer one, pruning techniques and training a smaller model to imitate a larger one, etc.) (e.g., neural networks) (NN), is claimed by the above researchers to support lowered computational power and subsequent actionable energy saving (minimization) without compromising any AI-led energy system performance across repeated energy setting experimental conditions.

Computation-Based Modes—CBM

The specifications energy field AI-programming can be executed in a diverse manner. That is, either distributed and (or) cloud computing one(s). In the first one, the above authors argue that the related training of the real energy solutions rests within several workloads energy nodes run, while in other cases, it can utilize multiple neighbouring energy assets. This type of computing tends to decrease processing timeline and diminishes energy collapse incidents (e.g., DeepSpeed by Microsoft) [3]. In another instance, federated learning can leverage energy workloads distribution contingent on occasion-based feasible, predictable and provision-available renewable energy resources to support a sustainable-driven energy sector system. In the second one, energy data reduction appears to be the regular procedure followed. However, centralized-based cloud computation seems to hold the benefit of heightened energy effectiveness from advanced physical and visible computing system components and central-directed spaces, maintenance, repairs, UQI, UQX and energy usage facilities. In that respect, therefore, the above researchers argue that the most computational architecture and energy-led effective energy domain modelling designs may be different from solution (ecosystem) to solution, resting on the existing energy agents (properties) each time (e.g., computational- and transmitted-based energy field data) that are more pertinent to the whole (inclusive) energy expenditure. Finally, other open-source, pre-trained and fine-tuned AI models (e.g., ModelScope by Alibaba) [4, p. 4] that support ongoing model-based learning, seem to assist energy savings by preventing from creating dissipated (or wasted) energy through repeating AI-model energy scene training.

Extended Reality—XR

Human Visual Processing—HVP

Wang et al. [5] and Razwan et al. [6] related work indicates that XR processing can accomplish multiple-times higher speeding and greater percentage of byte decrease. In a similar vein, the prior authors report that the same computation power can also be integrated into video compressing to limit (moderate) the energy consumption needed for transmitting energy-led data. Since the human eye bears restricted capacity to grasp elevated regional and time-relating detailed information or data, instead of processing all videos at fully-fledged quality, if the clarity of every other video framing is restricted, the achieved and (or) viewed resolution might be still sustained. Multifold video processing may also decrease both computation-based workloads and energy data transfer, thereby contributing to energy saving. Last but not least, any reduction in screen-brightening can further save a considerable proportion of energy, while maintaining a similar "glow" energy-based data viewing.

Blockchain—B

Energy Effective Consensus Mechanisms—EECM

In analogous lines, the delegated PoS development [7] configurations can run to authenticate energy exchange and integrate new blocks on the energy platforms. On the other hand, the proof-of-authority (PoA) (as above) structure tends to constitute the pre-approval energy participant validation or else, the consensus-reaching out processing, as that of lowered computational power-demanding and one of increased energy efficiency. The Ethereum energy sector B, in particular, denotes an indicative example of considerable energy consumption decrease [8].

Functional Optimization—FO

Decreasing the amount of energy-driven nodes and workloads during processing might support the reduction of unused functions and corresponding energy expenditure. Dividing the blockchains into smaller autonomous parts and directing any processing towards only one section, instead of the whole energy-led infrastructure, can limit the amount of idle functions and related energy consumption. Running a second-tier at the top of the foundation-based energy-linked B seems to reduce the workloads and the sophistication of the energy area transactions. For instance, the off-the-chain (OtC) compensation system (e.g., in Bitcoin, Ethereum, etc.) [9, 10] is considered to develop an immediate route between two energy sector members and ensure prompt, multimodal energy data transfer, providing that the overall number of such transfers will not exceed the energy data deposits [11]. In other instances, the "knowledge-zero rollups" (KZR, e.g., [12, p. 414] tend to perform the energy domain transactions off-the-chain, in a simple authentication proof generation that is sent back to the B. Despite the trade-off triggered by the generation-led proof, the above KZRs are considered to lower energy demands [13]. In a similar energy-efficient B update [14], a rather limited number of other than energy industrial sectors (e.g., telecoms, data centers, etc.) have invested in sustainable-driven energy processing. Energy-efficient (green or greener) AI-empowered solutions and B platforms are adopted within the energy landscape at a scarce range. XR ones, on the other hand, seem to still rely on energy system performance (computational power) rather than energy-friendly optimization (green intelligent systems).

2.4 Tentative Future Challenging Issues and Research Strands for Green(er) Intelligent Systems on Energy-Sector Metaverse

The aforementioned authors' evidence-based insights are equally reflected upon their additional suggested challenges and enquiry issues that pertain to the next frontiers, green smart intelligence technologies adopted on the energy sector Metaverse, as next.

User Quality Interface and Experience—UQI and UQX–Green Knowledge Performance Indicators—GKPIs

Most energy sustainability-led indicators seem to rest within the usual quality of energy system performance (QESP) elements. The envisioned Metaverse will run newer multimodal use cases energy data processing. In this context, before the design of intelligent systems comes into place to optimize the energy setting UQI and UQX green measurements, it tends to be rather critical to first fine-tune the quality experience (QX) paradigm that maps into UQI and UQX, respectively.

Metaverse "Green" Sustainability-Facilitated Energy Sector List

Eco-friendly sustainability strategy seems to be foundational for Metaverse energy suppliers to design a development energy route, for energy consumers to select green-based energy assets and for energy producers and (or) energy authorities to improve eco-sustainable, energy industry-led regulatory bodies and processes. Nevertheless, the acute number of smart intelligent technologies and energy investors or founders engaged, turns it as rather if not sweeping, at least most exigent to measure the total energy domain sustainability. Despite the individual assessment of currently advanced energy solutions that employ multi-perspective energy data measurements, their contribution to the Metaverse energy ecosystem and integrated UQX seem to be, yet, blurred. Actionable energy landscape models and applications are expected to be demanded in the near future, in order to embed cross-over technological energy domain assessment tools and lead the whole energy field evaluation.

Trade-Offs Between Energy-Driven Metaverse System Performance and "Green" Energy Consumption—EMSPGEC

Whilst rare existing work envisages on maximizing Metaverse energy system performance (e.g., keeping the highest precision or speediest AI-empowered models) [15], the previous authors argue that Metaverse energy field solutions might not request such high energy performance standards. In this envisioned scenario, the energy system performance could be somewhat given up to decrease either the carbon dioxide, methane, or energy per energy asset exchange. Nevertheless, any forecasted trade-offs between energy domain system performance and energy expenditure should assume future peak RIEI and public engagement studies.

Powerful Enticement to Embrace Green Intelligent Technology and System-Based Energy Modelling—GITSEM

Green intelligent building technology has been successfully implemented in recent years. One of the next smart technology generation challenging issue would be whether and how motivation and engagement in covering the emerging smart energy technologies of Metaverse might be fully supported. Cross-over enabling energy systems are envisioned as feasible to be interrelated across Metaverse. Scaling one smart energy technology over the other could trigger the extended utilization of other interrelated systems (e.g., sensor and video compression can decrease the transmitted energy data, but heighten the computational power-driven workloads). Across these lines, any Metaverse energy scene system architecture might use greater energy-saver dynamics than building upon autonomous sub-energy data networks (e.g., synergy-driven sensing, processing, recall and interaction). In that respect, the optimal system design thinking (i.e., open innovation, outflow to inflow innovation) appears to be crucial to steward future smart intelligent innovations towards the Green(er) Metaverse Energy Ecosystem (GMEE).

In drawing an overall summary of the above authors' framed insights on the envisioned, future, smart-intelligent, energy sector Metaverse running, they first reviewed a tentative sketch of the energy-"greedy" Metaverse smart intelligent associated technologies. Then, they envisaged on the latest technological innovations feasible to be aligned with the GMEE. Following-up on semantics-enabling Metaverse developments, they concluded their forecasting arguments with outlining the dynamics of some future emerging challenges that intertwine Smart Intelligent Systems with Green Metaverse energy-field infrastructure. Their aforepresented evidence-based reasoning is summarized in the following Fig. 2.2 that draws our reflective open innovation-driven, syn-integrative trade-off future REI and community engagement challenges between the immersive intelligent systems and energy-efficient (eco-friendly) Metaverse energy ecosystem.

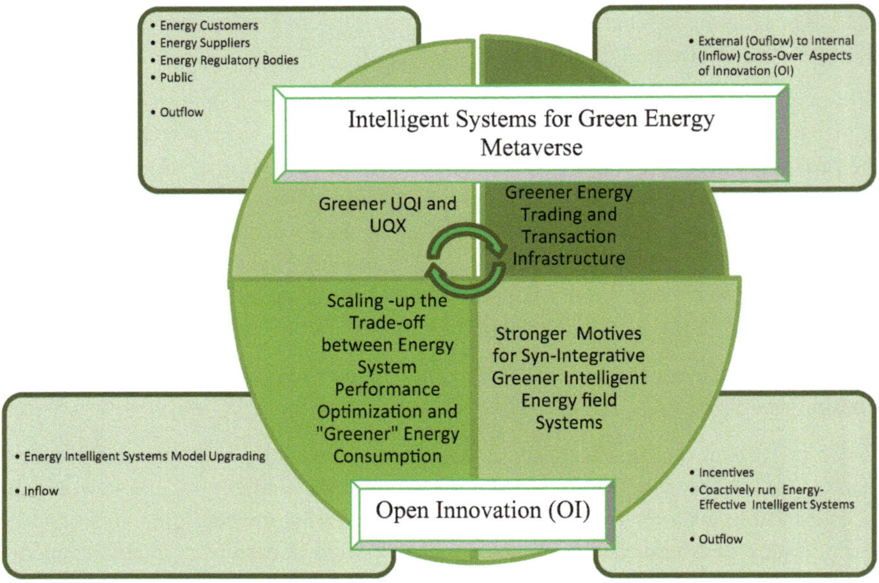

Fig. 2.2 Open innovation challenges reflecting intelligent systems and green(er) energy domain Metaverse

2.5 Smart Intelligence Energy Metaverse

Along the above argumentation "mirroring" the smart intelligence-led properties or particulars of the energy area Metaverse ecosystem, recent scholars [16, 17] indicate that Metaverse-driven immersive innovations (e.g., AI, VR, IoT, DT, B, etc.) bear the potential to radically transform the global energy-led domain by facilitating a distributed (decentralized), effective and sustainable-powered network. On one hand, this forecasted facilitation is considered to be a multi-level one, ranging from projected, context-specific energy research, development, innovation and enterprise, to sustainable power generation, repository, running and dissemination. On the other hand, it is thought to entail multiple challenging issues [11]. These extend from safety energy concerns, heightened energy spending, regulation energy reservations and lowered digital energy solutions adoption, to those of enhanced energy expenditure challenges, among others [18]. In order to resolve these challenging points authors in [11, 17, 19, 20] suggest a number of systematic actions (or procedures) that might endorse advanced cybersafety energy regulatory frameworks, optimized power utilization and improved guiding principles that could promote the energy-sector Metaverse smart intelligence adoption. In addition, the above authors recommend functional implementations, stressing the acute demand to perform additional scholarship, research and development on how best to optimize the integration of energy sustainable Metaverse tools into the energy-powered domain for further scalable efficiency. For example, they argue that AI is considered to optimize energy field expenditure and deployment, lowering resource energy misuse and promoting energy running sustainability by reducing greenhouse gas emission [17, 20, 21]. This smart-led energy facilitation might considerably lower the dependence on organic fuel (e.g., oil, coal, natural gas, etc.), supporting the conversion to a lowered carbon-dependent energy trading marketplace. Going further, authors in [11, 22] envision that B innovations are aligned with decentralized, safer, joint and interoperable energy sector solutions. In particular, B-distributed and visible commercial energy applications promote the adoption of renewable-led energy resources, supporting a more dispersed and viable energy trade marketplace. By decreasing trickery and confirming visibility in energy exchange, B innovations spread the basis for a more trustworthy and environment-friendly energy infrastructure. Moreover, the preceding authors tend to envisage that VRs can be an immersive and sustainable solution that aligns augmented and mixed reality (AR and MR) with generated extended reality (XR), lowering the demand for physical-based objects and in-house assessment, therefore, reducing resource-led expenditure and greenhouse gas discharges, promoting widespread ecological viability. Additionally, they endorse that the IoT innovations enable interaction and unseamed accessibility of the Metaverse network to the actual settings and physical surroundings, by linking objects (or assets) with tangible information. In that event, the IoT applications tend to advance the whole energy effectiveness through actual energy screening and optimized energy usage, lowering power and energy waste. This capability not only heightens the functional energy domain Metaverse trustworthiness, but also raises a sustainable energy-scene utilisation. Further, they stress that the aforementioned Metaverse immersive technologies (i.e., AI, VR, IoT, B, DT, etc.) work together to offer a powerful route for the advancement of sustainable and energy-effective Metaverse per se. In a nutshell, AI can enable smart intelligent systems, B applications are able to safeguard energy information safety and shared data, VR is capable of designing immersive energy innovations and IoT is expected to relate the actual energy-field settings to Metaverse. By connecting the above interrelated technological advancements, Metaverse energy-sector infrastructure can be facilitated and offer energy-interested individuals, teams, organisations, governments, enterprises and public agencies with immersive-driven, engaging and multimodal, person-led, digital-energy innovation surroundings.

Energy-Led Marketplace and Its Technological Transformation—EMTT

Digital energy innovations have transformed the global energy economy as that of a smart one. Through immersive applications (or systems) like sensor technologies, IoT and big data adoption, energy businesses can screen energy generation and supply chain management in actual timeline, optimizing the effectiveness of energy utilization, decreasing power expenses and developing the trustworthiness and security of the umbrella Metaverse energy network. In terms of the embedded Metaverse energy marketplace, the above authors argue that digital energy innovations transformation has raised the accurate preparation and combined research and innovation within the energy industry, building upon the renewable and decentralized power resources, thus enabling the digital fine-tuning function of coordinated energy supply chain management, enhancing the accessibility within the energy landscape and improving the efficiency of energy marketplace operation. As regards the energy usage, the digital innovations adoption enhances the personalized energy use. Through developments like smart cities and smart energy generation, storage and distribution networks, citizens, communities or agencies are able to engage in smarter energy-consuming activities to sustain smarter energy screening and management. Along this line of reasoning, authors in [21] project the utilization of Metaverse as a digital innovation energy market infrastructure, bearing the capacity and agency to support advanced and reliable energy trading and diffusion. Having said that, Metaverse appears to be further able to offer energy enterprises with a digital assessment context for the improvement of their energy facilities. This opportunity tends to decrease research and development costs and associated challenges and spreads the expansion of similar digital innovations. Metaverse can also provide energy customers with tailored assistance on their energy solutions, therefore, engaging them to endorse a tangible agency across energy-sustainable actions. Notwithstanding the significance of digital innovations in enhancing the worldwide energy sector, the above authors forecast or envisage the adoption of Metaverse-based energy solutions in the next decades.

Energy Digital Transformation and Metaverse—EDTM

Energy digital innovations propose the adoption of immersive solutions into all sections of the energy sector or energy enterprises, renovating conventional energy networks into those of more effective, accommodating and active digital energy ecosystems. This renovation incorporates the implementation of innovations such as AI, IoT, B, DT and smart data to support the optimization of energy production, expenditure, storage and dissemination. The end result target of energy innovations is to advance the emergence of sustainable, trustworthy and effective energy networks whilst decreasing expenses and greenhouse discharges. In this setting, Metaverse is a critical agent of extending digital innovations, mapping into a remote platform that integrates physical and virtual environments, providing a network where energy customers or suppliers are able to communicate through energy tools and with each other. Within the domain of energy technological innovations, Metaverse generates newer openings for energy infrastructure simulations, management and optimization. It permits the design of DT (i.e., the remote mock-ups of physical objects) that can be employed to screen and improve real-time energy sector networks. To advance the understanding of foundational Metaverse innovations in the energy industry, the following section outlines their particular application and the related ecological and monetary gains. In specific, the next part showcases the ways that AI, B, VR, IoT and DT are not only critical to sustainably efficient Metaverse platform per se, but also foundational in improving a power-effective, global energy domain.

Foundational Metaverse Smart Innovations and Their Merit in the Energy Domain

Resting on the aforementioned authors predictions (e.g., [17, 19, 21]), we might reflect upon their following forecasted insights, as next. First, the AI smart energy expenditure can be supported by advanced analytics and optimized power dissemination. Likewise, smart AI chatbots might facilitate the energy running simulation(s). As an end result, the above innovations tend to decrease energy misuse by improving resource distribution and utilization of renewable resources. In that respect, functional energy expenses can be either balanced or reduced, the energy infrastructure improved and the energy-led decisions more customizable. Second, B innovations appear to create distributed energy networks and safer energy exchange applications, ensuring energy data visibility and reliability. Further, they tend to endorse the inclusion of renewable energy resources, the smart grid customization, the prevention of energy misuse, the decrease of energy trading expenses, the advancement of energy trading visibility and the fairness of energy valuation. Third, the VR technologies seem to foster the learning and development (L&D) of energy experts and the creation of energy networks co-design (i.e., open innovation) and assessment. They tend to lower the utilization of physical resources and decrease the greenhouse emissions by substituting the physical assets. In addition, they might facilitate the energy L&D budgets, scale up the energy design procedures and promote energy sustainability. Forth and last, the IoT appears to support the real-world energy screening and running of energy applications, by permitting the scalable incorporation of physical and virtual objects. Moreover, it seems to foster the energy sustainability through the anthology of real-timeline energy data by decreasing potential energy information misuse. Lastly, it can lower conservation fees, advance functional trustworthiness and provide with optimized energy resource handling.

Interpreting the aforementioned authors, Metaverse seems to be not only the critical player that leads the technological renovation (renewal) of the energy domain, but also as the next frontiers agent. In acting as the latter, it tends to promote the worldwide sustainable development target by advancing energy effectiveness, decreasing energy resource-led misuse and fostering the adoption of renewable-based resource(s) energy. The implementation of such smart innovation(s) showcases the range and depth of Metaverse important dynamics in accomplishing a lower or zero carbon-dioxide emission energy market and enhancing energy viability. This supports our reflective, open, innovative-driven approach for current and (or) next years' energy handling. Focusing on the energy business-based merit of Metaverse, in particular, and analyzing the prior authors' reflections, we envision their insights on the following sub-sections, as next.

Developed Energy Visualized Data and Decisions Made—EVDDM

Metaverse is expected to provide improved visualization-related solutions that can support energy enterprises screen and examine sophisticated databases autonomously and interrelatedly. For example, workers can monitor the visualization of the whole energy grid operation, locate inadequacies and proceed with educated decision-making to support energy run optimization. This capacity aims to decrease the timeline and resource toolkit demanded for technical issues and safeguarding, hence developing the holistic energy network effectiveness.

Virtual Reality-Induced Employee Energy Learning and Development—VRELD

Metaverse facilitates immersive, diverse VR energy-cases simulated settings, like energy risk management responses and (or) energy device functioning in risky energy sector contexts. This category of L&D seems to improve the capacities of workers, lower the potential of energy industrial errors or disasters, and enables workforce greater preparedness to manage real-time energy-challenging issues.

Energy Sector System-Based Simulations—ESSS

Leveraging on Metaverse, the energy domain businesses might design (e.g., co-design, open innovate) and assess VR-led replicas of newer energy innovations or networks prior to applying them in actual energy settings. This is considered to be able to diminish the potential threat connected to disseminating advanced applications and grow (open) innovation by permitting speedy and iteratively based optimized energy system designs (or co-designs).

Virtual Reality Energy Factories and Decentralized Energy Properties—VREFDEP

Metaverse bears the potential to support the implementation of decentralized energy assets into VR energy plants, which are collections of lower-scaled energy production entities that can be monitored and operate as a mere energy unit. Resting on Metaverse, such VR energy factories might be tailored to weight energy supply chain management by regulating energy delivery and needs, increasing the steadiness of the broader energy grid and fostering the conversion to renewable-led energy resources.

Energy User-to-User Exchange—EUUE

The B innovations on Metaverse are considered to permit safe and visible customer-to-customer (C2C) energy buy-and-sell networks. These systems seem to provide energy clients with the opportunity to trade energy straightforwardly with each other, supporting more effective usage of renewable energy and decreasing dependence on conventional energy suppliers. Having elaborated on the aforementioned energy-led benefits of Metaverse energy smart management, the above authors continue with sketching first and analyzing next, specific subject and (or) research areas where Metaverse might be applicable within the energy domain, as follows.

Energy Research and Development—ERD

Metaverse can support the creation (i.e., co-design, open innovation) of energy services, promoting the deployment of newer energy digital innovations, their quality UI and UX design and increasing the energy industry operational effectiveness.

Energy Generation and Running—EGR

Metaverse technologies are able to provide the energy sector with sophisticated support on energy production, operations, L&D optimizing processes, advancing energy product or service features and promoting virtual monitoring and real-timeline energy market analysis.

Energy Repository and Dissemination (ERD)

Metaverse supports simulated and optimized innovations for energy warehouse and distribution facilities, allowing for effective and sustainable-led energy stewardship.

Energy Marketplace—EM

Metaverse can facilitate client-to-client (C2C) energy exchange, preset energy arrangements, visible energy trading, and the development of a powered energy market.

Energy Usage and Administration—EUA

Metaverse can offer energy-personalized handling, effective consumer experience and retention, energy trading and use of external to internal resources (i.e., open innovation) across the energy sector.

Energy Implementation Challenging Issues: EICI

The fusion of Metaverse and energy sector requires managing scalability, digital, monetary, utilisation, security, monitoring, regulatory and L&D demanding topics. A more detailed account of the above authors on the prior Metaverse implementation range concerns projected within the envisioned Metaverse energy sector is reported as next.

Energy Research and Development—ERD

Metaverse is considered to facilitate the research, development, innovation, enterprise and impact creation (i.e., co-creation, open innovation) of energy smart operations. An indicative example is gigantic energy enterprises, where VR functions contribute to the optimisation of energy services research and innovation (R&I) procedure. In such cases, real-world simulations are designed with energy engineers and software developers acting together and assessing them in a remote energy context. In that respect, they tend to showcase their capacity to define and solve emergent energy challenges before energy generation, enhancing energy R and I development effectiveness and decreasing expenses. Other energy giant industries seem to enable 3D-virtual setting software that offers the capacity to design and walk around immersive, 3D-remote energy spaces. This type of energy digital innovations is thought to be foundational in developing the generation of power or energy factories (e.g., [23]). This agency is projected by permitting energy engineers and software developers to envision and navigate with comprehensive energy modelling, thereby, diminishing lengthy energy production timelines and excessive budgets. The core mode that Metaverse might boost the development of newer energy design is forecasted by implementing DT innovations to perform remote energy assessments. Energy enterprises might speed up the development of newer digital energy innovations to enhance their comprehension of energy output and potential obstacles. This remote experimenting strategy could permit appropriate and time-saving energy data exchange against traditional energy testing methods. Seemingly, Metaverse might empower energy businesses to assess and improve energy preparation arrangements, in this way decreasing the dynamics for errors in energy design. Metaverse could also assist energy engineers and software developers across the energy design-based level, as well. For instance, they might efficiently interact with other energy experts in communicating energy development applying VR for 3D models and visualizations. Furthermore, as immersed through energy remote spaces they might collaborate with energy group associates or communities in actual-timeline, thus, accelerating the energy development and improving the energy design excellence. In addition, Metaverse could envisage energy production blocks, frontend-and-backend energy design and device-driven energy prototyping, which might assist decreasing energy development blunders and amendments. Throughout the generation level of an energy ecosystem, Metaverse might be able to provide with energy design simulations and visualizations including running of energy research resource requirements, timeline arrangements and energy risk-induced management. As such, it could allow for energy industries to better manage their energy design plans and energy quality-driven screening, thereby, feed-forwarding the energy blueprint performance. In parallel, through VR (remote) L&D, Metaverse might also improve the know-how energy stage and safety resilience, thus, lowering energy disturbances. In general, Metaverse is forecasted to strengthen the domains of energy R, I and E, scheduling and design of energy ecosystems, fostering the conversion and development of the worldwide energy sector into a more sustainable, eco-friendly and cost-efficient infrastructure.

Energy Generation and Running—EGR

Analyzing further the prior authors' insights, we can identify the next notes, as follows. Metaverse smart innovations bear powerful dynamics in VR energy generation and functions to support accelerated and remote energy supply chain management for energy enterprises. Well-known energy industries adopt VR to facilitate oil-ring-based growth. Resting on virtual showcasing, they might be able to increase the precision and effectiveness of their explored energy demonstrations. Moreover, they can use VR tools to direct earth scientists towards VR geographic expeditions. Such smart solutions tend to supplement professional arrangements and authorize them to get ready for acute conditions like dangerous energy leaks or explosions. VR intelligent innovations can transform employee L&D in energy businesses. Through real-context use cases and functional energy-driven procedures in a virtual setting, workers can practice their attributes in energy device functioning, advancing their energy security agenda, thus, lowering the timeline invested in energy safeguarding arrangements. What's more, VR smart innovations might also offer simulation-led L&D for energy crisis management, so that the energy personnel is able to address urgent energy conditions and proceed with informed decision-making.

Energy Repository and Dissemination—ERD

Along the same lines, Metaverse can offer varied smart simulations and optimized tools for sustainable energy expenditure and delivery. Relevant research has revealed that no substantial energy corporation has still provided with multi-dimensional simulations and optimized case studies applying Metaverse intelligent innovations for energy depository and allocation networks. Nonetheless, with the growing progress of immersive technologies nowadays and ahead, it is envisioned that more businesses are expected to implement Metaverse smart innovations to enhance energy expenditure, delivery and dissemination infrastructure for effective and trustworthy energy handling. For example, giant tech companies are showcased to use industry-led Metaverse solutions to support energy repository and distribution platforms. These systems endorse innovations like Azure IoT to assist energy enterprises effectively in handling energy dissemination and decreasing functional fees [24]. As regards energy expenditure, by implementing VR intelligent innovations, energy industries might be able to use multiple simulated and analyzed energy storage equipment, such as connected cells and gas repositories [25, 26].

Reflecting upon the above authors, across remote contexts, energy industries might simulate processes and KPIs like functioning settings, loading and refilling procedures, and effectiveness of energy expenditure equipment, performing sophisticated output and viability testing. By mimicking, energy suppliers could choose the most appropriate energy widgets to maintain skilled and sustainable energy agency. Concerning energy diffusion and delivery, Metaverse might provide VR energy simulations and optimized energy broadcasting networks. Energy businesses in a remote setting can lead diverse energy allocation and dissemination use cases, including energy delivery grids and infrastructure. Resting on VR smart innovations, energy industries can work with optimized stack development, electrical energy discrepancies, energy expenditure, and other allocation and distribution-networked energy KPIs. The optimized energy expenditure and dissemination plans could advance the energy delivery effectiveness, lower the energy misuse, and enable the overall energy infrastructure viability. By implementing VR intelligent solutions in energy expenditure, delivery, and dissemination, Metaverse can offer newer projected modes of optimized and decision-driven services for the energy sector. Despite rather lacking application industry examples so far, envisioning the above authors we note that in the next decades, energy enterprises can leverage VR smart settings to be enabled with energy simulated, diverse use cases, lending support to sustainable-led energy warehouse and delivery.

Energy Marketplace—EM

Moving forward to analyzing the prior authors, we project that as regards energy market and trade, it seems that Metaverse bears a dynamic capability. Through B and virtual payment innovations, Metaverse can facilitate CtC (peer-to-peer) energy exchange and the development of energy markets, speeding up newer options for energy deals. The preceding authors' insights direct towards forecasting that the worldwide energy B marketplace is considered to be able to expand notably across a number of smart energy innovation-tech countries in the following years. Such technological advancements can enable customers to be involved with CtC energy trading with other peers through B and smart contracts. As an indicative practical example, it might allow for users, groups, communities and (or) companies to allocate their solar energy and distribute surplus energy to other customers, supporting a cost-effective usage of energy. At this level, consumers could disseminate energy data on the energy supply chain management lifecycle, with other customers buying or selling energy resting on their requirements. This distributed energy exchange framework can be further applied within renewable energy community (es), where group affiliates exchange energy data and are actively engaged in CtC client energy trading to fulfil their demands more effectively. This CtC trading process tends to decrease energy dealing fee(s) and enhances competition in the energy domain [27]. When it comes to smart contracts, in particular, this type of innovation can also safeguard fair and visible trading, as well (e.g., [28–31]). The above authors seem to suggest that through smart contracts, Metaverse tends to empower automated energy deals and visible energy exchanges. Next, the smart contract technology engages customers in computerized energy trading, ensuring sophisticated and safe energy exchange. In such a mode, a central energy marketplace that fulfils energy delivery and power needs, can be developed on Metaverse. In this context, energy producers can offer data about their energy supply chain and clients can choose the preferred energy origin, resting on their requests. In addition, the promotion of energy wholesale centers can further improve not only energy trading, but also the interaction between energy supply chain agents, as well. In a nutshell, the preceding authors envision that Metaverse can provide consumers with energy proficient and eco-friendly options, speeding up newer development gateways of energy enterprises and transforming the mode that current global traditional energy trading is made.

Energy Usage and Administration—EUA

Metaverse bears the dynamics to support energy consumers in developing an engaged energy supply chain management life-cycle, providing energy businesses with an optimized energy ecosystem. Despite energy data visualizations and virtual energy societies, the adoption of Metaverse in the energy sector is still either at mere infancy or at completely early initiation. Nevertheless, related literature reflects on a scarce number of businesses having employed Metaverse enterprise intelligence within connected energy operations, yet stressing the opportunities for growth, sustainable energy management and associated issues to consider, as next (e.g., [32, 33]). As regards energy management, in specific, Metaverse can monitor energy usage in real-timeline through a virtual user interface (VUI), transforming the way consumers run their energy usage. Customers are able to use energy equipment through the VUI and visualize the corresponding energy utilization. Additionally, Metaverse can screen energy devices through smart house innovations, where residents control the use of their energy needed. Metaverse is expected to be further used as the infrastructure for interaction between energy suppliers (industries) and energy clients. The first could communicate their energy product or service development on Metaverse, while the latter can exchange their energy usage best practices and relative R&I updates. In that respect, a joint Metaverse energy business model can be developed and sustained. On another route, building energy networks within Metaverse infrastructure, can endorse collaboration between energy enterprises, customers and other energy investors (or founders, etc.), eventually leading the revolutionary conversion of the worldwide traditional energy industry.

Energy Implementation Demanding Topics—EIDT

The control and running of Metaverse within the energy domain tends to be still in its primitive consideration, with diverse energy coordination and safety challenging issues. A long-term adoption of smart innovations within the energy sector needs to be equipped to handle such enduring risks, efficiently. An outline of the latter is reported as next [11, 32, 34].

Ethics-and-Regulatory Conduct, Privacy and Safety—ERCPS

Hard energy data safety planning, running and assessment need to be developed to safeguard the safe and legitimate usage of energy agent(s) data. Consumer information and data safety is critical within the Metaverse energy enterprises where excessive gathering of data, expenditure and dealing out is needed. Actions such as encrypted data processing, authentication and exposure moderation measures need to be applied to avoid energy data breach and cyber attacks. Visibility and liability approaches have to be advanced to confirm energy consumer(s) reliability. Open access energy data gathering and usage, comprehensive authorized steps, exchanging energy client preferences and means to address swiftly energy safety risks, need to be established.

Smart Innovations Challenging Issues—SICI

Metaverse in the energy sector seems to experience a number of technological challenges. Most profoundly, these concerns align with the Metaverse smart and data integration. That is, Metaverse requests robust computational intelligence, high-standard information and communication technology properties and extended data adoption to achieve real-timeline interpretation. Building a steady information-related strategy, ensuring open access data options and accomplishing seamless-led amalgamation between varied mode solutions, can mitigate such challenges. Precise algorithms and modelling should be designed to reach the target of optimized power on the Metaverse energy platform. Such computational intelligence tools and models should process excessive, sophisticated energy data and decide over real-timeline energy conditions. Coding efficient algorithms and systems that provide with optimized energy networks and screening mass fine-tuning by bringing together big data, AI and ML, is considered to be supportive for actual smart energy solutions. Scalable models and standard data sharing are expected to foster the successful endeavour of energy Metaverse. The latter is filled with energy devices, sensing equipment and energy supply chain management approaches, with scalability, interconnectivity and data exchange between them, being a challenging issue. Therefore, it seems to be rather imperative to design not only data standardizations, but also energy domain standardizations that can feed into a proficient landscape for the implementation of Metaverse.

Budgets—B

The expenses required to build the energy Metaverse can be significant, resting on diverse factors, such as the range of energy device arrangements, the sophistication of energy modelling tools, and the stage of expert knowledge needed. Next, efficient cost-based energy monitoring is also critical to set up a sustainable energy enterprise model. Additional review and screening of connected energy expenses can leverage on the following points, as next. First, a meticulous assessment of digital innovations and devices in a comparison-based mode. That is, looking for properties with elevated energy efficiency, considering at the same time, whether the technology is scalable for further fine-tuning and optimization. Second, pulling together current network and tools through sharing resources and VR solutions, limiting duplicate investments and misuse of energy. Third, examining newer enterprise models like those applied on cloud service and distributed running mode. The aforementioned strategies are considered to be able to decrease outspoken energy venture and widen expenses to a larger team of energy agents.

Energy Expenditure—EE

Reflecting upon an expanded, interrelated VR infrastructure, the "Metaverse Energy Intelligent Management" is expected to request a massive amount of energy supply chain management to maintain its network, such as energy data or information and communication centers which could grow energy expenditure and discharges. In that respect, therefore, there seems to be the requirement to support an optimized Metaverse energy platform and ensure eco-friendly energy smart innovations to lower any environmental effects. The preceding authors recommend the below actions to trigger and sustain an "optimized-verse" energy eco-system, as follows. First, energy expenditure may be restricted by developing energy effectiveness in energy dataverse services. For example, lowering temperature technology, effective hosting connections, and efficient energy running and screening are considered to reduce energy misuse. Second, growing the range of environment-friendly energy on "Energyverse" by extending the use of renewable energy resources (e.g., solar electricity, eolic energy, hydropower, geothermic energy, etc.) in the energy supply chain management. This would limit dependency on conventional, non-renewable energy resources and greenhouse gas releases. Third, logical allocation and utilization of energy storage sources in the "Energyverse" infrastructure. This could be achieved through energy loaded balance, joint-resource smart innovations, prevention of surplus expenditure of particular digital energy assets, and decrease of energy misuse.

Individual Energy Usage—IEU

Given that the energy industry starts preparing for the integration of Metaverse, the actual challenging issue rests within the limited individual attributes or the required competencies across authorities to consider and (or) efficiently implement "Energyverse" intelligent innovations. Growing employee tech-based and smart innovations practice to support their adaptation to the swiftly developing requirements of the "Energyverse", would be a valuable venture. Along this line, the developed technological attributes would enable them to endorse the energy Metaverse fast emerging and modified requirements, safeguarding energy enterprise(s) sustainable development. This could be achieved through business, academic and R&I collaborative energy-led programmes such as by integrating cross-discipline energy awareness with handy implementation, relevant information and communication technology

(ICT) networks, developing talent management, designing or showcasing the most recent smart energy innovations, and exercising related problem-solving skills. On another route, the provision of ongoing energy-induced L&D courses could assist energy leaders and energy workforce in mastering the newest trending intelligent energy innovations. In particular, such energy L&D programmes could entail the application of "Energyverse" smart innovations, optimized energy networks and smart running to satisfy the demands of energy experts, globally. In addition, investment in establishing energy Metaverse labs and simulated settings to provide interested energy stakeholders with appropriate use cases in experimental energy business conditions, should also be followed. Such smart laboratories are forecasted to be able to offer simulated, everyday energy infrastructure and support them to innovate smart energy data analysis, remote energy modelling skillsets, and wider energy infrastructure optimization capacity.

In sketching the dynamics and challenging issues pertinent to the adoption of Metaverse smart innovations within the energy sector, the above authors emphasize the potential to advance the inquiry, effectiveness, and viability of "Smartverse" energy infrastructure, opening novel energy enterprise ventures. These outcomes tend to align with current related work that supports the range and depth that smart innovations bear for the establishment of intelligent energy networks (e.g., [35–37]). Nevertheless, all prior authors seem to project the directions or management of smart energy challenging issues like energy safety and energy cost-and-sustainability-effectiveness, which appear as foundational for the umbrella integration of Metaverse intelligent innovations in the energy industries. Reflecting upon the aforementioned authors' insights of core research, development, innovation and enterprise (RDIE), we extend them by embedding these into our "Energyverse" envisioned landscape visualized in the following Fig. 2.3 flowchart, as next.

As all the preceding authors forecast, the adoption of Metaverse smart innovations tends to describe important wins for advancing "Energyverse" sustainable development and running. Our presented outreach not only commensurates with, but also expands prior conventional research and development, recommending next generation, smart multimodal routes for Metaverse trading and authorization within the energy industry. As regards practice-based implementations, the current work seems to forecast the best strands to integrate Metaverse smart innovations within the energy domain and thereby, to optimize intelligent energy procedures like energy generation, design, expenditure, distribution, and research and innovation. This proposed Metaverse-driven framework appears to demonstrate that featured smart energy simulations are able to greatly decrease energy expenses and grow energy resource usage. In addition, whilst Metaverse smart innovations have started to be implemented in other businesses, the present work indicates its exclusive envisioned projection within the energy sector, offering a basis for improved R&I. The latter is targeted at tackling energy adoption challenging issues and scaling these innovations across diverse energy angles. In order to promote the growth of our envisioned "Smart Energyverse" and its business implementation in the energy arena, building on or extending the prior authors' reflective insights, we suggest the below system design thinking

Fig. 2.3 "Energyverse" smart adoption and sustainability challenges

(i.e., open innovation) "Smart Energyverse" envisaged research and development implementation, as next.

Optimized "Smart Energyverse" Networks—OSEN

All the above authors elaborate on additional R and I needed to highlight the optimization of energy plans, functions and running of energy infrastructure built upon Metaverse intelligent innovations, as projected next. For example, research and practice should advance the Metaverse smart simulations (or modelling) to act as replicas of the actual energy networks. Further it is necessary to adopt AI, ML and data science to support smart energy optimized deployment. These attempts target at advancing the energy effectiveness and limit energy expenditure, offering obvious advantages regarding related energy costs and ecological energy outcomes. Along the aforementioned authors' lines, investing in R and I projects aligned with energy enterprises and research organizations, might support design (i.e., co-design, open innovation) and assess such "Energyverse" "smart" simulations, successfully. The emphasis could be directed towards, for instance, regional areas where energy optimization would drive to baseline, transparent energy effectiveness benefits. Securing bursaries from cross-over financial resources or investors to promote the design and dissemination of such future energy modelling, including also generative AI-led one, might also help guaranteeing critical and next-generation resources for efficient and scalable energy smart implementation.

Renewable-Resource Energy Implementation—RREI

A renewable smart resource energy solution could be implemented by exploring the dynamics for greater-range inclusion and business intelligence transmission of renewable energy smart resources across Metaverse. Setting a remote energy marketplace to support the optimisation of renewable intelligent energy production and expenditure, in conjunction with predictive analytics, could advance the maintenance of sustainable energy grids, supporting a modification towards more eco-friendly energy systems [38, 39]. Across the prototyped "Smart Energyverse" market, the renewable energy could be exchanged in real-timeline, with B innovations enabling safe and visible energy trading. Such projects can be projected to be piloted in areas with increased renewable energy generation and screen their impact on power grid-based sustainability.

Energy Digital Twin Modelling—EDTM

Campana et al. [40] and Bhutta et al. [41], among others, postulate that this kind of modelling could employ smart Metaverse innovations supported by real-timeline information and communication energy services. The latter might provide with energy decision-making assistance, energy error handling and forecasting energy functionality, bearing the potential to enhance the scalability, transparency, trustworthiness, security and effectiveness of smart energy networks. In a first instance, the DT innovations could be implemented to foundational network(s), such as energy power factories or big renewable farmhouses to help energy performance optimization and reduction of energy functional interruption (e.g., [42, 43]). The latter would be further improved through iterative and scalable data integration on "Energyverse".

Energy Regulatory Framework Generation—ERFG

Reflecting upon the above authors, supporting diverse academia, research and enterprise to plan and develop energy regulatory frameworks and innovation-led standardization, is expected to scale energy application, scalability, and enable settlement with energy security and privacy-data acts. Establishing global energy stakeholder associations to deliver business regulations for smart energy Metaverse innovation-driven implementation and managing challenging issues like energy data confidentiality and safety, authentication and regulatory clarifications, could foster the secure and efficient distribution of "Energyverse" smart solutions. In a nutshell, all preceding authors conclude that given the transformative dynamics of Metaverse smart innovations in the energy enterprise, critical topics request additional envisioned research and development. Digital innovation challenges and cost-effectiveness are considered to be foundational for the fully-fledged realization of the impact these smart energy innovations could bring forward. Overall, they highlight that the Metaverse smart innovations that endorse DT, AI and ML simulated intelligence and distributed energy running, bear the dynamics to critically improve the effectiveness, venture capital and sustainability of such "next frontiers" energy networks. Most importantly, these "Smartverse" innovations fuel the assimilation of renewable energy, contributing to the restriction of greenhouse discharges and the promotion of smart and sustainable energy operational optimization. Nevertheless, challenging issues like energy data privacy, elevated energy expenditure concerns and energy management I and L and D appear to be pertinent. Handling these concerns through improved energy cybersafety, energy cost-efficiency and human capital upskilling, seems to be critical

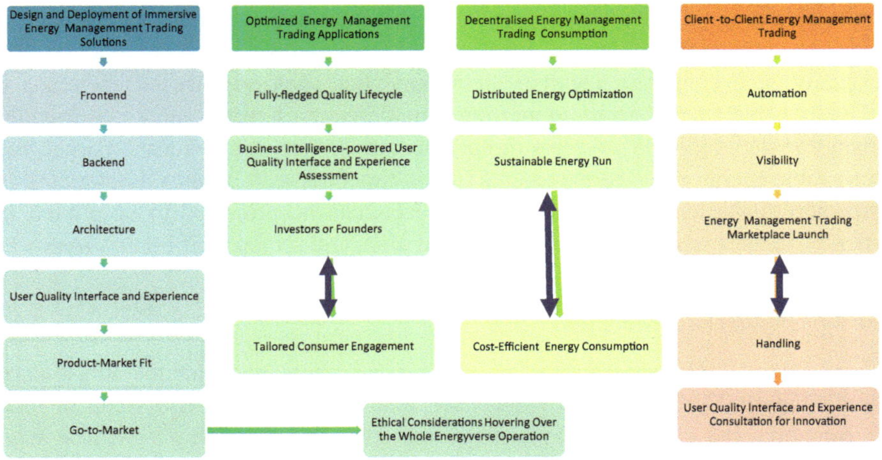

Fig. 2.4 The projected Energyverse management trading open innovation properties and challenges

for effective "Smartverse" energy management optimization at a worldwide scale. However, additional RDII appear to be required to fuel the fully-fledged dynamics of these smart Metaverse innovations and address the existing energy sector concerns or risks. Hovering over and extending the preceding authors' reflective insights by depicting our above conceptual model envisioning the "Energyverse" forecasted management ecosystem (i.e., Fig. 2.3), we might expand this projected model as an open innovation ecosystem that integrates external (i.e., outflow) and internal (i.e., inflow) dimensions of smart Metaverse energy management "mirroring" the sub-level of energy trading management landscape innovation aspects and interrelated challenges, as follows (i.e., Fig. 2.4).

Reflecting upon diverse authors' insights for Metaverse energy supply chain infrastructure envisioned interventions and management (e.g., [37, 44–46]), we note the following narratives (scenarios), as next. A B platform infrastructure can be introduced to the setting, where no other active agent needs to be present to promote energy delivery. The administrator of the distributed energy supply chain is envisioned as a data structure link (i.e., node) in the network that can execute the duplicate exchange between the energy investor (or founder) and the client (or consumer). Each data structure link can validate the supply provision and administer the documentation of the exchange to all other network data structures. The energy expenditure on the local Energyverse supply-driven marketplace can be fully performed by the B platform. Along this line, by introducing a DT and replicating the ecosystem utilized by clients on the Metaverse-led platform, the agency of the purchaser can be lowered. In this respect, as the replicated infrastructure on Energyverse single-handedly can spot the demand for energy and its sum, it can transmit this call to the consumer on the B platform and buy energy at the lowered feasible cost. Adding to the above, upon the agency of the energy vendor, by properly simulating the generation ecosystem on the Metaverse-driven platform and the range of energy generated at whatever time, data on the quantity of energy that can be disseminated from Metaverse-led infrastructure is offered to B and minimizes the agency (or capacity) of the supplier in performing this responsibility. As a last note, the supplier only generates and the client only spends, while in-between, all energy transactions are made by replicas on Metaverse space and the support offered from the B platform. On a similar strand envisioned by diverse authors (e.g., [17, 47–49]), a business procedure and script modelling of local energy supply chain management (or intervention) on the Metaverse appears to be projected. This ecosystem seems to consist of visualized and optimized multimodal procedures. These processes are integrated into the local energy Metaverse-led supply chain management, from energy production and warehouse to energy dissemination, exchange, expenditure, revenue and authorization regulation. A tentative example of the prior authors' projected model is outlined on Fig. 2.5, with a substitute of it illustrated on Fig. 4.6, as follows. The above proposed and alternate models are expanded as our unified, open innovation-driven model in the following Fig. 2.7. Additional arrays of similar authors' insights on local energy supply chain management visualized on Metaverse (LESCM) (e.g., [50–52]), offer a systems design thinking (i.e., open innovation) for this Metaverse-led optimized procedure. Their connected explorative models seem to envision the dynamics of the revolutionized, sustainable, local supply chain energy production, transmission, exchange, expenditure, revenue, maintenance and regulatory standards compliance on Metaverse. Figure 2.5 sketches a reflective outline of the prior authors' iterative-driven "universal" model of LESCM, as next. The first level (or stage) of the model builds upon the regulation-led compliance requirements. The next one integrates the properties of energy expenditure and management. The third aligns with the energy production and warehouse (maintenance) properties. The fourth maps into energy exchange (trading) with the fifth delving into that of energy transmission and the last one with that of associated energy revenue (budget).

Figure 2.6 depicts the prior sketched universal, alternate, envisioned structured model of the local energy supply chain life-cycle management on Metaverse. This visualized model (i.e., scheme, framework) tends to rest within an iterative design, as next. First, it starts with the regulation-based compliance. Second, it moves towards the optimized energy expenditure and management. Third, it goes forward with the virtual energy markets. Fourth, it continues with the integrated smart grid-and-disseminated energy resource(s) assimilation and fifth, it finishes with the generated smart contracts and B-led energy digital assets.

Last but not least, Fig. 2.7 represents our envisioned, unified, and iterative model that seeks to amalgamate all preceding authors' projected models of local energy supply chain management-Metaverse as an open innovation one, as next.

In all the three forecasted sustainable business model envisioned cases above, the aim is to offer an envisioned insight, knowledge and (or) reflection upon the ways Metaverse might transform the mode(s) in which energy generation, transmission, and consumption operate in an external (outflow) to internal (inflow) innovation iteration. As the aforementioned projected models showcase, a local energy supply chain on Metaverse can offer a disseminated, visible, scalable, safe, integrity-led, energy-efficient, and interoperable ecosystem for energy suppliers and clients to exchange

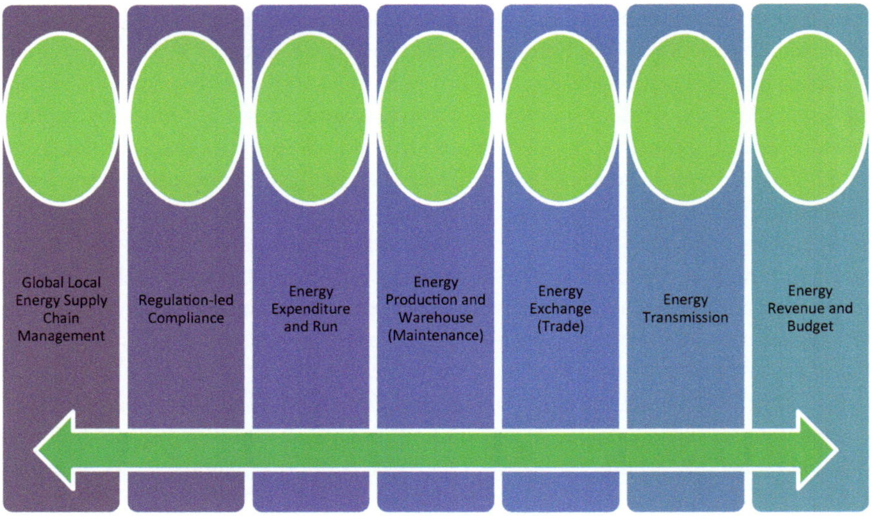

Fig. 2.5 Envisioned local energy supply chain management business model

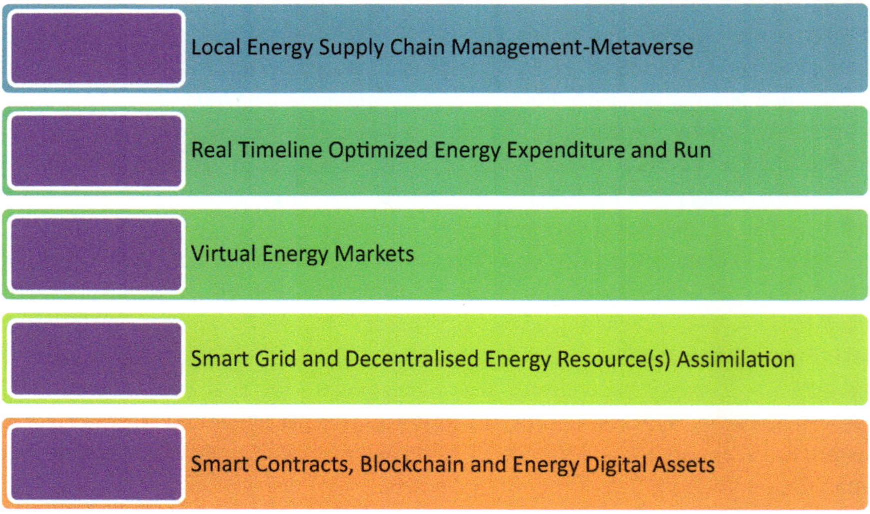

Fig. 2.6 Alternate LESCM-Metaverse envisioned sustainable business model

energy assets. Notwithstanding the dynamics of the above envisaged benefits, all prior authors also emphasize the potential challenging issues or concerns that need to be taken into perspective, as reflectively outlined in the following section(s).

Benefits and Challenging Issues—BCI

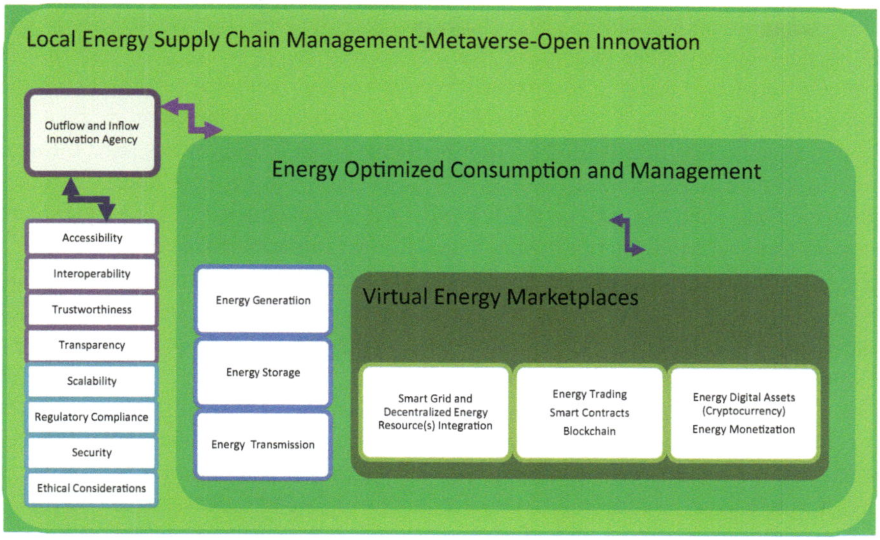

Fig. 2.7 Author LESCM-Metaverse open innovation sustainable business model

Diverse researchers (e.g., [31, 46, 48, 49, 53]) project the potential strength of the LESC on Metaverse-driven infrastructure being present in multiple domains. Reflecting upon the above authors' insights, it seems that the "Energyverse" offers a distinctive context for deep evaluation of innovative-led energy solutions through simulated modelling, permitting the examination of varied context and supporting the growth of supply chain energy business models on Metaverse. In addition, it appears that the advanced screening of the energy expenditure configuration and schemes within the Metaverse can offer greater data insight or overview and foster improved energy effectiveness and precise energy expenditure prediction. Further, the disseminated Metaverse ecosystem tends to serve straight line CtC exchange(s), overcoming the request for intermediates, such as traditional energy services or stakeholders. This feature seems to constitute Metaverse as a perfect implementation workspace for local energy trading. Adding to this, the enhanced assimilation across energy devices, services and suppliers together with energy investors (or founders), clients, agencies and regulatory bodies, seem to support direct overview for validated decisions and greater efficiency along energy expenditure run.

On another route, the prior authors envision that the challenges of the LESCM on Metaverse seem to endorse a foundational venture needed to set up the established ecosystem and the digital innovation service required for the energy marketplace DT infrastructure. In addition, the intricacy of assimilating multimodal suppliers, investors, founders, consumers, agencies, inter alia with varied objectives, motives and concerns, tends to present a challenging issue, whatsoever. Associated technical expertise and interests might also restrict stakeholder engagement, while the evolving immersive AI, GenAI, and ML technology could prevent broader implementation across the LESCM ecosystem. Last but not least, a potentially restricted client database on Metaverse could challenge interested energy investors (founders or agencies or services) to achieve ROI (or economies of scale) (ES) and run on a monetized way. In going forward with the preceding authors' reflections on the potential benefits of the LESCM on Metaverse, the following notes might be articulated, as next. First, the digital energy innovations can be deeply assessed through simulated narratives, testbeds, A and B version (AB) testing, user quality interaction (interface) and experience (UQI and UQX) evaluation, etc. This element can support more scalable energy business modelling. On the other hand, the advanced screening of energy expenditure properties, schemes, and actions is expected to end up in improved data overviews that foster energy effectiveness and prediction of energy expenditure. Further, the Metaverse disseminated network seems to facilitate a straight-line, CtC energy exchange, by overcoming the request(s) for intermediates like traditional energy services and thus, constituting it as an optimal application for the LESCM. The enhanced assimilation of energy devices and energy stakeholders including energy service suppliers, clients, agencies and energy regulatory bodies, seem to serve a more sustainable energy supply chain management on Metaverse. Additionally, the automated energy infrastructure tends to foster greater energy effectiveness and reduced energy budgets, while also serving the 2050 global zero-carbon optimal goal. Adding to the above, the usage of the infrastructure seems to direct at lowering the energy consumption by restricting potential dependence on a centralized-driven energy network (or resources), thus, advancing the viability and the budget-friendly feasibility of energy for everyone. On another strand, Metaverse offers a UQI and UQX-friendly context, facilitating all interested stakeholders to engage in the "Energyverse" marketplace. Moreover, the utilization of Metaverse-driven immersive innovations appears to safeguard the legitimacy and trustworthiness of the interactions due to its fail-safe security features and optimal transparency-led ecosystem. Notwithstanding the above comments, the Achille's heel of LESCM on Metaverse can be summarized, as next. First, the request for monetized venture to design the ecosystem and immersive technologies requires serving energy marketplace DT engineering innovations. Second, the intricacy of joining together diverse contributors with varied expertise, interests, and objectives. Third, the digital innovations know-how might restrict engagement. The Metaverse-led digital innovations or emerging technologies remain quite fresh, not broadly implemented, which tends to reduce their adoption on the Metaverse-LESCM. Fourth, the non-extensive consumer database on Metaverse seems to challenge energy suppliers (e.g., investors, founders, agencies, services, etc.) to reach cost-efficiency as the scale of operations expand.

Prospects and Risks—PR

The Metaverse appears to bear gigantic dynamics to transform the global energy systems design by advancing its transparency, interoperability, scalability, safety, economy of scale and energy-efficiency, inter alia. By diminishing intermediates in conventional energy grids, the Metaverse can revolutionize universal energy as more transparent, reachable, and inexpensive for everyone. It also fuels the growth of large-scale, innovative, and open innovative, in particular, energy sustainable business models that can leverage dexterity and suppleness in worldwide energy marketplaces.

In addition, the Metaverse offers a perspective or a route to tackle foundational challenges such as universal energy accessibility and deficiency, especially in emergent regions or states. What's more, extended data distribution among engaged contributors within Metaverse seems to exaggerate improved information and communication agency among energy marketplace stakeholders, directing towards advanced energy effectiveness and decision-making.

Building upon the above arguments, rationale, and insights, it could be argued that there are a number of risks that need to be taken into consideration or reflection, as follows. First, the DT ecosystem which is foundational for the operation of Metaverse, might experience underfinancing due to a short-lived emphasis on energy ventures, affecting its longer-term sustainable status. Second, risks associated with information technology security are raised among the broader population, which could impact the reliability and pervasive integration of Metaverse-led energy marketplaces. Third, the adoption of B digital innovation in energy marketplaces within Metaverse, might address limitations or ethical considerations due to ambiguous regulatory frameworks. Fourth, the dynamics of the energy sector with continuous digital innovations could trigger restrictions and challenges related to scalability, interoperability, integration, trustworthiness, consistency, and conformance concerns within the LESCM on Metaverse.

Reflecting on and extending the aforementioned authors' perspectives of the LESCM on Metaverse, some additional insights could be envisaged, as next. First, Metaverse bears the dynamics to radically transform the LESCM, setting energy more trustworthy and open access for everyone, limiting at the same time the expenses of energy through the restriction in intermediates across conventional energy production and transmission networks. Second, the empowering of immersive, digital innovation business models is actionable and applicable across swift and adaptable energy marketplaces. Third, the sustainable prospect of Metaverse to resolve challenging issues around energy accessibility, shortage and (or) surplus, cross-regionally. Fourth, the capacity of Metaverse to advance shared data contribution, engagement, communication, and cooperation among global energy marketplace stakeholders. Fifth, the Metaverse ecosystem is empowered to support the generation, maintenance, delivery, and dissemination of renewable energy sustainable resources in a scalable, transparent, secure, and cost-efficient mode, promoting universal zero-carbon transmission and affordable consumer energy budgets. Related to the above argument, as customers become more knowledgeable of the wins of B digital innovation(s) and distributed energy marketplaces, the request for such ecosystem(s) is expected to grow.

On another hand, following-up on the preceding authors' reflective insights regarding the noted risks of the LESCM on Metaverse, these might be collectively envisioned, as below. First, the DT innovation ecosystem tends to experience decreased long-term investment, due to a constricted emphasis on shorter-term energy-driven funding. Second, the escalating array of global concerns about broader data confidentiality, ethical considerations, regulation, management, security, and accessibility on Metaverse infrastructure. Third, the application of B technology in energy marketplaces could be decreased or trigger ethical concerns due to the ambiguous regulatory frameworks that existed. Fourth, the global energy sector is continuously growing and developments in digital innovations might cause restrictions in the LESCM on Metaverse (e.g., compatibleness, inter alia).

In a final overview, reflecting upon the prior researchers' reflective insights in outlining the promising agency (and capacity) that Metaverse bears in smart cities-driven sustainable energy supply chain management, the below points might be proposed, as follows. First, empowered by the notion of revolutionizing the globe into a more sustainable environment, analogous to the inspiring vision of Metaverse, their articulated work explores how this envisioned operationalization might promote sustainable energy transmission within regional management. By reporting the dynamics of Metaverse in this sector, the present review highlights the potential of metropolitan cities (or regions) to become "Meta-verse" cities. Building on the capacity of Metaverse-driven immersive innovations in transdisciplinary sectors, these could accelerate procedures, set aside time and effort-investment, and save expenses.

In a nutshell, the related literature presented above tends to emphasize that the energy transmission in the LESCM on Metaverse can be distributed (i.e., not centralized), thus, reducing the request (or demand) for intermediates. Integrating DT innovations and simulated solutions on Metaverse, the energy vendor(s) agency will be limited, as the ecosystem by itself will screen energy requests (or demands) and buy energy assets at the cheapest costs. In a similar vein, energy suppliers can offer an optimized energy transmission by promoting the simulated generation infrastructure on Metaverse, advancing streamlined processes and restricting their participation (or contribution). The assimilation of Metaverse into "smart city-led energy supply chain management" bears a promising perspective for effective and sustainable-driven LESCM. By integrating DT innovations, intelligent recommender ecosystems and optimized simulations, smart cities or smart urban development can endeavour towards a more environment-friendly future that hierarchizes effectiveness, customer-centered strategies, and environment-led sustainable network. As a whole, the inclusive (integrated) insights review presented seems to highlight the request for additional R, I, E, engagement, and strategy growth and practice. This would extensively reap the potential merit of Metaverse outflow to inflow (i.e., open innovation) sustainable energy supply chain management and vice versa, universally.

Along an analogous route of intelligent energy exchange management, authors in [54] discuss that nations globally shift from traditional, coal, gas, and oil-based energy networks to sustainable, renewable-led energy infrastructure. To efficiently adopt renewable-based energy resources, multimodal energy stream and screening are requested. Conducive to this multi-dimensional energy provision and monitoring, user-to-user (U2U) or C2C energy trade is adopted. In order to facilitate a secure and visible U2U (or C2C) energy trading network, a safe information and communication route along with a data storage space are requested. In their current work, they investigate the adoption of B innovations into a renewable energy grid. They attempt to highlight four foundational domains. First, the energy trade, second, the eco-friendly dihydrogen supply chain management, the energy need-reply programme, and the tracing of renewable-based energy accreditations. Additionally, the prior authors explore the ways B technology might handle the challenging issues in these sectors and prevail over these barriers to establish a distributed, energy-driven infrastructure. The major aim of their present work is to offer a comprehension of the manners in which B can play a triggering role in unravelling a multimodal energy run, eventually transforming the mode energy is produced, organized, used, and distributed.

Implementation of the Blockchain Technology into the Energy Domain—IBTED

B innovations bear the dynamics to revolutionize the route energy or power is produced, disseminated, and used, due to its distributed and visible properties. They can further validate and verify the dealings safely and address all critical issues dealt within the energy domain, such as the adoption of renewable resources, effective energy trade, grid running, and plenty others. They are able to support C2C energy trade without any glitch and hence, assist incorporating renewable resources into the grid, successfully. B technology could be further applied in newer energy innovations such as clean hydrogen. It can be utilized in the renewable hydrogen supply chain management and (or) be connected to clean hydrogen charge points in the next years. It could also be employed to amalgamate IoT equipment and electric cars (ECs). By adopting B innovations, the energy industry bears the capacity to grow as an added, sustainable-based and distributed energy network.

Energy Trade (Exchange)—ET

A foundational application of B innovations in the energy management is the distributed (i.e., decentralized) energy trade. The above authors argue that, conventionally, in the traditional power management grid (see Fig. 2.8), energy management trading appears to be a central procedure with financial institutions and energy enterprises serving as mediators, supported by the possible existence of "pro-customers" ("pro-clients"). The latter are individuals or agencies (entities) that both generate and consume energy, obscuring the conventional lines between customer (consumer) and producer. The suppliers are the centralized authorities associated with the power transmission in this trading mode. These businesses are supplied with power from the central grid and move it to the customer, without being able to estimate the range and depth of neighbouring needs and supply chain use cases. The financial institutions are the centralized authorities for all financial exchanges, taking a vast amount of time to validate these dealings. This action elevates the rate of power, and in most cases, the mediators are not able to know about the neighbouring needs and (or) supply chain management use cases. On the other hand, the researchers propose the C2C B energy trading management scheme where C2C energy exchange is connected to the grid and monetized (i.e., payments made) in a decentralized, flow-driven and secure iterative mode (see Fig. 2.9 next),

Varied B websites with various settlement rules can be utilized to foster visible and safe dealings across the grid, as B technology largely augments safety, visibility, and effectiveness. As regards energy trade in "microgrid" (pp. 1, 3 above), the central scenario of B technology is C2C transactions, which ensure immediate trading between clients and peers aimed to buy very high-quality energy. This procedure overcomes central mediators like services and financial institutions, thus, decreasing

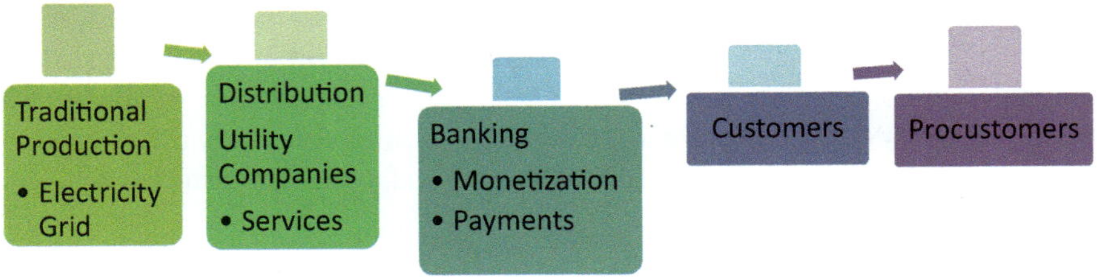

Fig. 2.8 Traditional energy exchange management

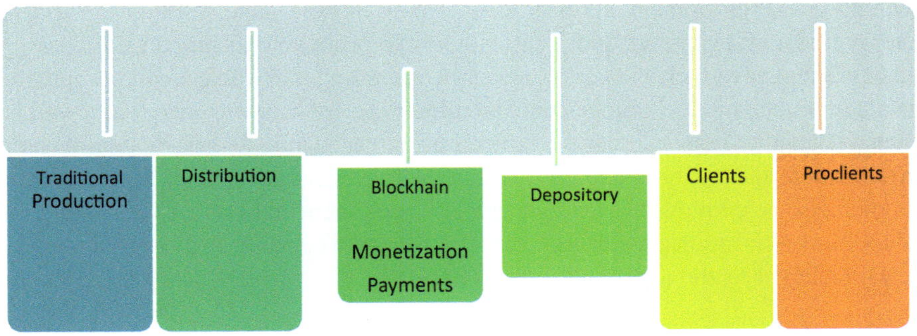

Fig. 2.9 Blockchain client-to-client energy management trade

exchange fees, improving autonomous energy trade, possibly upgrading the energy utilization of infrastructure customers and further advancing the total effectiveness of the network. It can also enhance the adoption of renewable resources by permitting more immediate and neighbouring exchange of "renewable energy certificates within microgrid" (p. 14 above). The aforementioned authors note the variety of enterprises and websites for energy trade adopting B technology across the world. Nevertheless, they claim that there seem to be diverse technological and legal-based challenging issues related to the adoption of B innovation. As regards C2C energy trade, B technology appears to indicate important barriers, the most evident of which tend to be the scalability-associated ones, particularly with "Ethereum" (p. 4 above), preventing its expanded integration. Given the increased trading transactions, a number of B platforms could manifest challenges in keeping up with satisfactory running speed and ability. As a result, this can end up in exchange delay and growing cost. Such development tends not to be appropriate for C2C energy trade, where concurrent interaction and energy transition have to take place. Moreover, the vast computing power needed for the energy expenditure within B sets the latter as rather less sustainable and challenges the mere aim of the C2C energy trade. In addition, lowered interoperable B running and less seamless network interaction, set the stage for a less shared trading. The energy marketplace is equally disintegrated, with decentralized operation schemes and marketplace configuration in every distinct, lacking the capacity to nurture an extended, interrelated C2C energy trading network. Concerns over regulations that appear not to critically take into account the separate features of distributed energy trading websites, come into effect. Resting on the above rationale, the prior authors argue that the complicated procedure of adopting the B innovation into structured energy networks seems to call for foundational changes to traditional modelling. In addition, the maturity of customer self-reliance and the improvement of comprehension concerning distributed network(s) can impose barriers to consumer integration, learning, and development.

In the next years, scalable and latent energy trading can be developed through systems like "PoS" and layer-2 scalable applications (pp. 4–5 above) that appear to advance the effectiveness of B infrastructure. Designing well-developed regulatory-related structures is vital for the upcoming integration of B technology within C2C energy transactions. Such structures or modelling need to adapt to the distributed origin of these solutions. Enterprise-led, standardized regulatory frameworks are expected to enhance an interoperable, interconnected and incorporated, distributed energy trading context. Growing smart contracts can exercise a critical effect by offering functional features like programmed settlement resolution, valuation, and provisional deals to expand the client know-how in C2C energy, fostering awareness of B innovation(s) and supporting confidence among end-consumers.

Clean Hydrogen Supply Chain Management—CHSCM

The continuous quest for green and sustainable energy tools has shifted the global focus on "hydrogen" (pp. 1, 12 above) which discharges zero-carbon emissions and particularly, clean hydrogen. This is generated by utilizing renewable resources and appears as a revolutionary answer to the worldwide climate disruption. In addition, its higher level energy concentration per item load in comparison with other fuels, seems to permit greater timeline energy warehouse, for instance, for cars. This advantage tends to extend its automotive-associated benefit towards empowering generators for renewable resource repository and substituting gas, oil, and coal in business operations. Moreover, its dynamics for household generation can increase energy safety by decreasing the dependence on introduced gas, oil, or coal, transforming seriously spoiling enterprises into "zero-carbon" (p. 11 above) discharge ones.

Piling up hydrogenium (hydrogen in Latin) for future utilization can be comparatively more manageable than that of electric power, providing with the dynamics to store at scale. In spite of its advantage, though, there are still challenging issues related to it. At present, generating clean hydrogen tends to be more costly than conventional sources of energy, due to the greater rate of renewable resources. Hydrogen-based re-charging, particularly for cars, seems to be at an infant stage, preventing excessive integration. The tech-associated concerns for hydrogen warehouse, delivery, and dispersion appear to require specific technological advancements and ecosystem growth. Nevertheless, acknowledging the importance of clean hydrogenium, funds are rolling. State-led policy, regulatory frameworks, and digital innovations tend to trigger and (or) boost this development. The global renewable resources society argues that in the next 25 years, clean hydrogen would be able to comply with the 25% of the worldwide energy need(s), stressing its dynamics to revolutionize our energy infrastructure (p. 11 above). A number of governments, such as the Indian one, for example, plan to generate millions of clean hydrogen in the following 5 years (each year) (p. above) enabling India to decrease carbon discharges by 50 methylcyclopentadienyl manganese tricarbonyl (MMT) and restrict bringing in non-renewable energy resources by more than 11.4 billion euro (p. above). Compared to its complemented fuels, clean hydrogenium appears to be really sustainable led. No remaining carbon discharges seem to be ejected when it is generated or used, setting it as a valuable asset in mitigating global climate disruption. Whilst expenses and generation restrictions are present, foundational attempts are on the go to overcome them. The decline in clean hydrogen fees to €1–€2/kg in the next 5 years sets a rather hopeful landscape (p. above). In addition, related funds are anticipated to attain €170 billions in the next 5 years (p. above) indicating the shared investment and responsibility to release the dynamics of this technological innovation.

In the energy sector, the clean hydrogenium (or hydrogen) seems to fuel a swirl, since it is acknowledged to be a critical agent in the transfer to sustainable-driven energy properties. The generation, usage, and delivery of this entity form a rather sophisticated supply chain management that can vastly leverage from the B digital innovation(s). The clean hydrogenium is generated through power produced by renewable resources (e.g., sun or wind energy) to split aqua into hydrogenium (or hydrogen) and O_2 (oxygen) (p. 13 above). This ecological and sustainable approach to generate hydrogenium tends to be inherent to the clean energy infrastructure. There seem to be diverse stages in the clean hydrogenium supply chain management, pertaining to generation, storage, delivery, and usage. The B digital innovation(s) with their distributed origin denote revolutionized ways out to overcome the challenging issues across the clean hydrogen supply chain management lifecycle. The advantages of utilizing B technology in the clean hydrogenium supply chain circle are described in detail, as next.

Marking Out and Visibility—MV

These properties seem to be two of the most critical advantages that B digital innovation(s) bear within the renewable resource hydrogen supply chain management continuum. From generation to delivery and dissemination, the B technology introduces a distributed, public-led "ledger" (p. 3 above) that records each operation and progress of clean hydrogenium. By sustaining visibility, each user (client) in the supply chain management cycle is provided with updated and accurate data relevant to the resource, merit and carbon dioxide discharges of the clean hydrogen. The adoption of the B digital innovation(s) permits consumers and suppliers to monitor the fully fledged pathways of the clean hydrogenium, thus, verifying its operational processes and safeguarding compliance with eco-friendly regulatory frameworks. In this context, they also support responsibility among clients and development of a trustworthy and sustainable energy provision.

Monitoring the Carbon Dioxide Pathway—MCDP

The clean hydrogenium is well-documented for its lowered carbon dioxide trace and eco-friendly generation when compared to conventional approaches of hydrogen-based creation. To trace this carbon dioxide pathway along the supply chain management and verify the related discharges, B innovation(s) appear to be crucial. They make the carbon dioxide tracing data readable for validation (or authentication), since each and every exchange in all levels is projected into an unassailable, ledger-based entity (pp. 10, 12, 14 above). Therefore, by setting up B technology, both authorities and customers are fostered to deliver informed decisions resting on trustworthy and educated information regarding the environmental-associated outcomes of clean hydrogenium.

Safe Data Shared and Networking—SDSN

In a sophisticated supply chain management circle with varied customers (or suppliers) and vast amount of information, exchanges have to be safeguarded and performed effectively. B technology supports safe data distribution, thus, maintaining security and privacy. This is deemed possible by yielding an authorized inspection of the B platform to every customer, facilitating them to access related data whilst defending confidential information. Thus, supporting data sharing and networking among consumers and suppliers, enhancing the effectiveness and interconnection of the clean hydrogenium supply chain circle. B innovation(s) also permit(s) seamless interaction among suppliers and customers, allowing them to share and transmit core data and (or) information, thus, optimizing procedures and advancing sustainable operation. The above authors note that there seem to exist varied applications (or implementation) of B technology in the clean hydrogenium sector, globally. Nevertheless, adopting the B technology opens the route to the development of C2C energy markets, enabling customers to trade immediately excessive energy. Through smart contracting, trade can be automatic, hence, enabling secure and visible energy transactions and restricting the demand for 3-way authentication-based mode. This kind of distributed strategy supports customers by offering them greater monitoring over their energy options and also facilitates the usage of renewable resource infrastructure (e.g., photovoltaics, wind generators, etc.). Resting on the trading of energy "tokens" (pp. 9–10 above), financial transactions can straightforwardly be validated by a smart contract and the demand for financial management can be limited, hence, providing cost-efficiency. Adopting smart metering, B technology offers real-timeline indicators for energy availability with each supplier and customer. The prior authors note multiple clean hydrogenium projects being underway or planned to be initiated in the next years using renewable resources (e.g., solar energy) across the globe. The latter are based on a B-driven smart contract ecosystem that is set to provide automated trading and ensure the monitoring of the hydrogenium supply chain management lifecycle, successfully.

The above authors argue that an unexpected rise in need for clean hydrogenium may damage the ecosystem, ending up in time-consuming trading and elevated cost. Incorporating B technology into the present supply chain management life-cycle requires foundational technical modifications. Such amendments might experience resistance from suppliers or consumers used to traditional methods. Nonetheless, standardized data formatting and protocol are needed to safeguard seamless transactions and interactive data running in the supplier chain cycle. Moreover, the energy expenditure of particular B infrastructure, specifically that utilizing the "PoW consensus-seeking operation" (pp. 13, 15 above), appears to be extensive and augments ecological alarms. This shortage of energy effectiveness does not seem to support the critical target of the clean hydrogenium domain and that of sustainable-driven run. In that respect, the preceding researchers call for further research and development to constitute the B innovation(s) to be more energy-led effective by utilizing different "consensus protocols" (p. 3 above). In addition, they also call for a robust and standardized clean hydrogenium accreditation procedure, globally, in order to guarantee the trustworthiness of the supplier chain management life-cycle and make straightforward the monitoring and authentication running.

Further to the aforementioned challenging issues, regulatory-based qualms could prevent the adoption of B technology into the clean hydrogenium supplier chain cycle. To address these issues, the above authors emphasize the need for additional research in the next years, thus, critically affecting the running of clean hydrogenium supply chain lifecycles. The interconnection of clean hydrogenium supplier chain circles within the overall energy ecosystem should be investigated to integrate clean hydrogenium into the energy domain and not merely as an energy source. Conducive to the above, B technology offers soft implementation among diverse renewable resources, aligning with a more integrated infrastructure for the organization and running of clean energy. In addition, IoT tools can also be connected to B technology to enhance screening and data gathering in actual timeline. This interconnection can be executed by incorporating smart sensing equipment that sends information and data in a straight line to the B technology, which advances the precision and pace of data for the generation, transmission, and expenditure of clean hydrogenium.

Actual-timeline Need Reply—ATNR

The need for electricity has increased as a result of residents' increase and digital innovations, which traditional electricity ecosystems cannot address by themselves. To lower this need-supply chain breach and produce power through renewable resources, "microgrids" (p. 9 above) are being applied. Still, this on its own seems not to be sufficient to close the enormous mismatches between need and supply chain and preserve grid robustness and thus, a demand-response (DR) mechanism is utilized. The DR appears to be a vital system that permits microgrids to adjust to energy usage optimization and advance grid stability, lower running cost, and heighten the network's performance. The DR aligns with mechanisms that prompt customers to shift their power expenditure schemes in reaction to grid circumstances or charging signs. In traditional infrastructure, services fine-tuned produced power to narrow or end the need-supplier break. Against the conventional approach, DR authorizes customers to turn into dynamic contributors in grid running. Consumers are able to select from different DR systems, such as cost-based DR and immediate (straight) load monitoring tools. In the straight load screening ecosystem, services can screen particular client equipment centred on customer permission. While in cost-related DR systems, restricted fees are provided along with low-season time to empower clients to transfer non-vital overloads (e.g., dishwasher or laundry machine operation) to these hours, and increased costs along top-need timelines to confirm that consumers decrease expenditure or utilize substitute energy resources (e.g., house batteries).

The authors claim that using the DR mechanism(s) in "microgrid(s)" (p. above) seems to provide multiple benefits, such as advancing the effectiveness of energy usage within the grid. It also decreases the reliance on costly and infecting energy factories by prompting customers to change expenditure outside of high demand times. It further assists customers and suppliers engaging in lower costs. Clients profit from decreased energy invoices during less-energy demand times, whilst the "microgrid" (p. 10 above) provider(s) benefit from limited functional budget by restricting dependence on costly, high-demand timed energy. DR schemes seem to critically support grid stability by dynamically handling need(s) variation. They also lower the percentage of power shortages, especially in "microgrids with higher integration of renewable resources" (p. 13 above). Nevertheless, the adoption of DR projects in the microgrid domain tends to manifest various concerns. The core critical challenging issue rests within safeguarding vigorous contribution from customers. Even if a number of clients are prompted to register at the DR scheme persuading their peers to alter their expenditure patterns that they are used to, it entails informed campaigns and engagement initiatives. In addition, garnering and handling customer energy data for DR objectives demands powerful cyber-safety actions and defined data rights modus operandi to guarantee customers' confidence. To efficiently handle real-timeline data and transmit suitable DR indicators, high-level information and communication ecosystem and refined modelling are needed.

The above concerns can be addressed by immersive innovations like B technology. By using B digital advancements, a safe and visible account of energy expenditure and related enticements can be piled up constantly, hence facilitating confidence and prompting wider customer contribution. This diminishes the demand for a central collection of customer data and also reinforces data safety and confidentiality. Employing B technology, smart contracts could also be used to lead DR automations, simplifying the whole procedure. Smart contracting can automate DR processes resting on predetermined settings, like cost variability or grid jam, thus limiting central-based monitoring and the demand for energy mediators. As B technology is unassailable, the data included in the B-driven infrastructure cannot be manipulated. This grows confidence in the procedure and protects from malevolent juggling around with sensitive data. Hence, the foundational features of B ecosystem like stability, visibility, and distributed running, considerably advance the effectiveness of DR schemes.

The above authors emphasize that despite B technology provides diverse benefits in DR projects, there appear to be numerous challenging issues connected to seamless-driven grid running and concurrent data transmission. PoW B platforms like "bitcoin and ethereum" (p. 4 above) make an enormous effort to manage vast amount of actual-timeline data related to DR schemes. The latter request the B technology to be incorporated into the grid seamlessly led, for which the B infrastructure has to align with the safety indicators, information and communication processes, and data categories utilized in the network. The present DR schemes and relevant conventions have to be tailored to adopt B technology tools and also to overcome concerns like settlements, data possession, and accountability within the B DR infrastructure. Although the aforementioned concerns tend to be critical ones, continuous research and innovation attempts are based on them to additionally investigate and redefine the adoption of B innovation(s) in actual timeline DR ecosystems. Scalable-led B modelling with high-level reaction time and decreased energy expenditure is being examined to develop the holistic effectiveness of the infrastructure.

Renewable-Resource Energy Accreditation—RREA

The speedy incorporation of renewable resources into energy grid(s) demands powerful tools to boost their growth and make sure the ecological KPIs related to their production are appropriately taken into consideration. The renewable energy certificates (RECs, p. 14 above) offer a way out of this by enabling tracking capacity and liability for each Megawatt per hour of energy produced by renewable resources. The accreditation procedure of RECs appears to be a structured strand. An assigned accreditation-driven agency utilizes deep meter-and-screening platforms and validates the properties of the energy supplied to the grid(s). Post-effective validation, one renewable resource certificate is provided for every Megawatt per hour of renewable resource produced. The REC possession is computer-based and enlisted and the specific renewable resource is expected to be connected to it. The RECs support the development of further income for engaged energy clients, thus motivating higher membership in renewable resource production. This, sequentially, develops the growth of newer renewable resource schemes and their financial feasibility. What's more, the RECs foster enterprises to align with the renewable resource acquisition responsibilities (RRARs) set by particular authorities. In a number of regions, enterprises and agencies are authorized to obtain a proportion of their energy from renewable resources. Even though the businesses may not comply with this obligation, the RRARs could be aligned by acquiring the same amount of RECs. Lastly, the RECs facilitate the maintenance of grid safety and steadiness, promoting the variety of energy production, hence decreasing the reliance on non-renewable energy resources. Notwithstanding their important benefits, the RECs can be sensitive to replication or scam and could likely threaten the ecological and financial merit related to REC possession. Unsafe RECs can direct towards a duplicate challenge, which amplifies the expected ecological advantages of the grid(s). This can wear away confidence among marketplace customers, leading to a decrease in the advancement of newer renewable resource schemes. In that respect, the safety and trustworthiness of RECs tend to be significant to ensure confidence and confirm the efficiency of the infrastructure. These challenging issues can be addressed by integrating B technology to safeguard REC distribution and running. The B innovation is an unassailable and decentralized, ledger-based ecosystem, which means that all infrastructure members keep a duplicate of their trading exchange(s), limiting the potential for scam. All ecosystem customers are able to validate that RECs are authentic with appropriate owners, with any possible manipulation made being simply identified due to the cryptographic link in the B platform. If any manipulation is made on a specific piece of data structure in the B, the linkage of that structure changes and this is indicated in the next array of data structure. This mapping mechanism of the B extensively decreases the threat of replication or tampering. On another level, B technology appears to enable visibility across the full REC circle. From introductory production, authentication, to following possession transmissions, all trading exchanges are notated on the B and can be validated at any time, hence improving confidence among marketplace clients. As a final point, by utilizing smart contracts (SCs), the accreditation procedure can be automatic, endorsing foundational running such as REC allocation and transaction(s). In this manner, a streamline of the administration operation, a restriction of trading cost(s) and the effectiveness of the REC marketplace are supported. The above authors continue their arguments by noting some other research that explores b-driven RECs for "microgrid tools" (pp. 5–9). In outlining this strand of multiple and multimodal research, Bhavana et al. [54] discuss that it utilizes SC on a B ecosystem to optimize the automatic distribution, transactions, and resolution of RECs. Further, they argue that B innovation can foster a more effective and trustworthy RECs marketplace by reducing mediators and restricting trading expenses. Moreover, a B-driven C2C energy transaction and REC ecosystem in "microgrid" (pp. 14–15 above) can also be suggested, allowing for uninterrupted energy and REC trading between the microgrid-associated consumers, fostering a distributed and visible energy market, ensuring safe REC exchange within the microgrid infrastructure. Moreover, the above authors stress the reliability of the B technology for REC operation on microgrid(s), safeguarding the veracity and validity of RECs by logging each exchange on the B ecosystem. All in all, the B innovation can improve confidence and make more efficient the REC running, adhering to a more trusted and latent microgrid-led infrastructure. In another research case, the above authors emphasize the integration of B innovation with smart metering and IoT equipment to validate and monitor REC production and expenditure. By utilizing B technology visibility and safety elements, the prior authors envisage a REC ecosystem that can precisely distribute and authenticate tokens within microgrid(s) platform. By utilizing a distributed and unassailable ledger-based infrastructure, the B technology is able to revolutionize the REC marketplace, allowing for microgrid-driven clients to participate in safe and C2C energy exchange, successfully.

Conducive to the above, the prior authors argue that the PoW B platforms seem to be rather not capable of managing vast amount of data-based information, concurrently, facing scalability-led challenges that can drive towards sudden rush in transmission responsiveness and restricted efficiency of the infrastructure. In cases where the data quantity advances, the weight on the B rises, thus, growing the timeline needed to perform a mere transmission, successfully. Bearing in mind, at the same time, that incorporating data from all ecosystem partners, like renewable energy resources (RER), facilities and supervisory authorities, tends to be rather challenging without any standard data formatting and information and communication modus operandi. In addition, there could also be replicas of RECs delivered due to the vast amounts of data existing in the infrastructure. The B ecosystem should be capable of avoiding this duplicate documentation, in order to safeguard the validity and power (dynamics) of the credentials. These challenging issues could be addressed by scalable-driven, "next frontiers", generated B(s) network(s). Notwithstanding the above, the authors emphasize that continuous research is set to deploy energy-effective, interoperable-led B energy management ecosystem(s). In that respect, the integration of B technological advancements in REC supervision provides breakthrough opportunities to optimize renewable energy usage in "microgrid(s)" (p. 3), as the global energy industry evolves into a more sustainable-led generation. Figure 2.10 outlines the authors' proposed model of combined benefits and challenging issues associated with each one of the aforementioned implementation(s) opportunities, as next.

Despite the multimodal challenging issues associated with the B infrastructure within the energy domain reported above, the above researchers claim that these can be addressed by the emerging B digital innovation(s) modified either through the development (coding) of several levels in the superior (higher) layer of them or through the transformation of the authentication procedure of their deals. The novel innovation Ethereum B with "PoS" (pp. 4–5 above) authentication of exchange(s) modus operandi contrasting the previous Ethereum one of "PoW" (p. 4 above), is proposed by the authors as tentative instance(s) of energy-efficient B solutions. The risks or concerns associated with the application of B innovation(s) might be resolved by designing "level-2 B tools" (pp. 3–4 above). These operate through "PoS algorithms" (pp. 4, 5 above) that tend to be more eco-friendly (i.e., energy-supported) effective. Ethereum platform appears to have integrated the PoS algorithm by now,

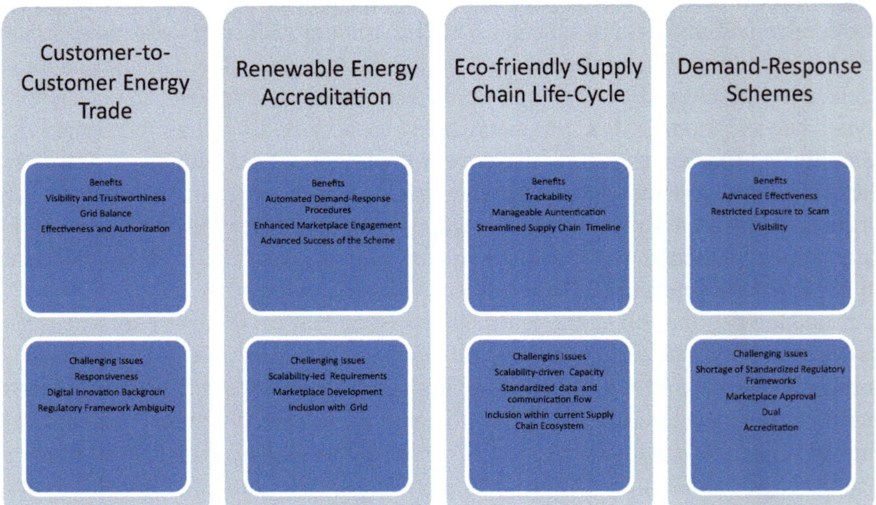

Fig. 2.10 Wins and barriers of the implementation of blockchain digital innovation within the energy-friendly management ecosystem

while a level-3 B for Ethereum is additionally designed to address the scalability-and-system responsiveness-based concerns. These emerging and recent Bs seem to be needed more in the energy ecosystem, due to the vast amount and range of data requests. In the coming years, the authors note that the challenging issues of operational and scale-driven agency and capacity across diverse B innovation(s), including the assimilation of B(s) into the present energy sector of economically developing regions, should be overcome to efficiently utilize such cutting-edge digital innovations to serve human capital and sustainable transformation. In addition, the adoption of demand-response schemes, C2C trade (exchanges), and renewable energy accreditation (certificates), are proposed by them as supporting the advancement of actionable viability of the energy marketplace. On government-based part, the above researchers stress the demand for regulatory frameworks to be mandated in order to integrate disruptive and seamless-led B innovations in the energy landscape.

The aforementioned authors conclude their current work by stressing that their overview offered a wide-ranging exploration of the prospective implementation and merit of integrating the B innovations in the energy marketplace, by emphasizing the critical revolution they impose. Further, they note their attempt to elaborate on how the visibility and stability (consistency) of Bs can safeguard data management vigorously. This effort is strengthened by their reasoning on how B can address the challenging issues met by the conventional energy marketplace, efficiently. Moreover, they argue that they have further highlighted the demand for the endorsement of B in foundational fields like customer-to-customer energy trade (exchanges), the eco-friendly hydrogenium supply chain life-cycle, the real timeline demand-response schemes and the renewable energy certificates. The researchers indicate that they also compare the current, evidence-based, project research and innovation (R&I) strands across each B implementation area, including C2C energy marketplace and the hydrogenium supply chain management workflow. On another hand, their present work has also investigated the benefits and concerns (or risks) related to the adoption of B innovations in each implementation area. The challenging issues adhering to the usage of B technology have also been highlighted in a detailed manner. In a similar vein, the current B platforms seem to experience varied risks (or challenging issues), including scalability-driven agency and capacity, system responsiveness (i.e, latency potential), ability to operate smoothly inbetween various B wokflows, energy eco-efficiency, and capacity to operate in-sync with regulatory frameworks that need to be revised for greater comprehension and compliance commitment. Furthermore, assimilating B(s) with the present energy ecosystem tends to be a big and challenging endeavour that has to be dealt with for an error-free, smooth running, and effective implementation of B innovations in the energy marketplaces. Overall, the above researchers have provided us with a helpful account of energy-friendly research and practice insights across different solutions and the dynamics of B digital innovations in the energy infrastructure, including renewable-led workflows. Expanding [54] operationalization and modelling of B innovation(s) in the energy-friendly sector, we might consider their paradigm as one that "mirrors" an open innovation-led framework, adhering to the exercise of external to internal (and vice versa) agents, workflows, networks, operations, benefits and challenges, in the eco-friendly, energy

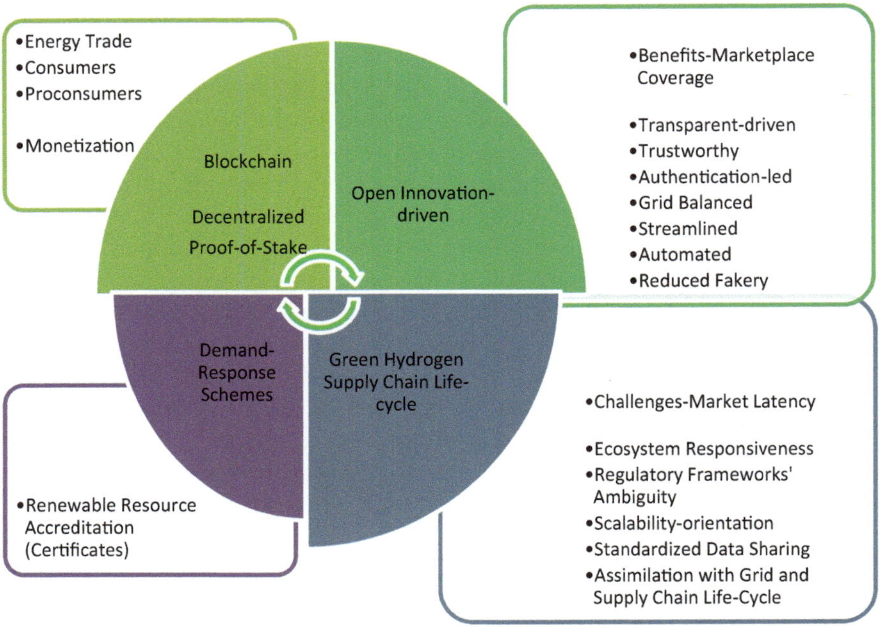

Fig. 2.11 Author open and eco-innovation-driven blockchain energy management ecosystem

B landscape. The following Fig. 2.11 depicts our outlined open innovation-led model within the B eco-friendly energy management sector, extending [54] ecosystem, as next.

2.6 Conclusion

The seldom envisaged future "SmartEnergyverse" management of open and eco-innovation ecosystem is envisioned to radically transform nearly any angle of smart intelligence and sensing energy marketplace network. It is envisioned to be built upon smart energy, sustainable and eco-generative, decentralized, scalable, secure, regulatory, accredited and monetized innovations intertwined around peer-to-peer energy consumers, suppliers, regulatory agencies, factories, and communities on Metaverse. This projected "Energyverse" smart business model management aligns with multimodal benefits and challenges echoing trade-offs between smart intelligence and energy sector Metaverse endorsed implementation, performance, services and agents. Such "next frontiers" envisaged ROI inquiry bears the dynamics to be efficiently followed in the near future.

References

1. Zhang S, Lim WYB, Ng WC, Xiong Z, Niyato D, Shen XS, Miao C (2023) Towards green Metaverse networking technologies. Adv Fut Direct 1–10. https://arxiv.org/abs/2211.03057
2. Rafique W, Qadir J (2024) Internet of everything meets the metaverse: bridging physical and virtual worlds with blockchain. Comput Sci Rev 54:100678. ISSN 1574-0137. https://doi.org/10.1016/j.cosrev.2024.100678
3. Rasley J, Rajbhandari S, Ruwase O, He Y (2020) DeepSpeed: system optimizations enable training deep learning models with over 100 billion parameters. In: Proceedings of the 26th ACM SIGKDD international conference on Knowledge Discovery & Data Mining (KDD'20). Association for Computing Machinery, New York, NY, pp 3505–3506. https://doi.org/10.1145/3394486.3406703
4. Zhang A (2024) Research on enterprise digital transformation: a case study of Alibaba. SHS Web Conf 208:01014. https://doi.org/10.1051/shsconf/202420801014
5. Wang H, Dong H, El Saddik A (2025) Immersive multimedia communication: state-of-the-art on extended reality streaming. ACM Trans Multimed Comput Commun Appl 21(7):Article 190: 1–33. https://doi.org/10.1145/3721292
6. Rezwan S, Wu H, Cabrera JA, Seeling P, Reisslein M, Fitzek FHP (2025) MLcXR+: multilevel semantic compression for 3D immersion over 5G networks. IEEE Access 13:164771–164786. https://doi.org/10.1109/ACCESS.2025.3610029
7. Asif R, Hassan SR (2023) Shaping the future of Ethereum: exploring energy consumption in proof-of-work and proof-of-stake consensus. Front Blockchain 6:1151724. https://doi.org/10.3389/fbloc.2023.1151724
8. Zhang Z (2025) The application of proof of stake: advantages, challenges and future development. ITM Web Conf 73:03013. https://doi.org/10.1051/itmconf/20257303013
9. Saini A et al (2025) Blockchain-powered Metaverse: enhancing security, decentralization, and interoperability for next-generation virtual ecosystems. AIJR Proc 7(6):75–81. https://doi.org/10.21467/proceedings.7.6.10
10. Karim MM, Qu Q, Cai Y, Liu T, Meng X (2025) Bitcoin reimagined: a comprehensive study of ordinals and inscriptions protocols for Web3 asset innovation. Blockchain Res Appl 100379. ISSN 2096-7209. https://doi.org/10.1016/j.bcra.2025.100379
11. Jayavarma A, Parakkat Kesava Panikker P, Nair MG (2025) Revolutionizing the energy sector: exploring diversified blockchain platforms for a sustainable future. Front Blockchain 8:1544770. https://doi.org/10.3389/fbloc.2025.1544770
12. Huynh-The T, Gadekallu TR, Wang W, Yenduri G, Ranaweera P, Pham QV, Benevides da Costa D, Liyanage M (2023) Blockchain for the metaverse: a review. Futur Gener Comput Syst 143:401–419. ISSN 0167-739X. https://doi.org/10.1016/j.future.2023.02.008
13. Sathio AA, Rind MM, Awan SA et al (2025) An energy-efficient and privacy-centric consensus protocol for secure Blockchain finance using zk-SNARKs. Res Square. https://doi.org/10.21203/rs.3.Rs-7175542/v1
14. Zhang Y, Zhou C (2025) Can blockchain technology promote green transformation? Evidence from Chinese Listed Enterprises. Systems 13:258. https://doi.org/10.3390/systems13040258
15. Al-kfairy M (2025) Greening the virtual: an interdisciplinary narrative review on the environmental sustainability of the Metaverse. Sustainability 17:7269. https://doi.org/10.3390/su17167269
16. De Giovanni P (2023) Sustainability of the Metaverse: a transition to industry 5.0. Sustainability 15:6079. https://doi.org/10.3390/su15076079
17. Raman R, Kautish P, Siddiqui A, Siddiqui M, Nedungadi P (2025) The role of metaverse technologies in energy systems towards sustainable development goals. Energy Rep 13:4459–4476. ISSN 2352-4847. https://doi.org/10.1016/j.egyr.2025.04.012

18. Zhou B, Xi C, Li C, Li G, Yang B (2023) Metaverse applications in power systems. JPGT 1(1):54–64. https://doi.org/10.61369/JPGT.0570
19. Han Y, He X, Zhang H, Zuo Y, Zhang W, Ai Q, Chen Z (2024) Metaverse framework designing for energy scheduling in energy internet of things considering emergence. Digital Twin 1:3. https://doi.org/10.12688/digitaltwin.17873.2
20. Ma Z (2023) Energy metaverse: the conceptual framework with a review of the state-of-the-art methods and technologies. Energy Inform 6(42). https://doi.org/10.1186/s42162-023-00297-w
21. Shimpo F, Sokołowski MM, Mikusek P, Chawla Y, Figueroa RB Jr (2026) Cybernetic avatars for energy transformation: regulating the digital revolution for enhancing sustainable attitudes of energy consumers. Digital Policy Regul Govern 28(1):35–51. https://doi.org/10.1108/DPRG-12-2024-0320
22. David LO, Nwulu NI, Aigbavboa CO, Adepoju A, Adepoju OO (2025) Evaluating the use of blockchain technology and identifying critical success factors for the successful implementation of renewable energy projects in sub-Saharan Africa. Int J Sustain Energy 44(1). https://doi.org/10.1080/14786451.2025.2449867
23. Leong WY (2025) 3D printing for renewable energy technologies. ASM Sci J 20(2):1–10. https://doi.org/10.32802/asmscj.2025.1978
24. Shanubhog SA (2025) AI-driven smart energy management in industrial facilities: leveraging azure cloud technologies for real-time optimization. J Comput Sci Technol Stud 7(8):611–616. https://doi.org/10.32996/jcsts.2025.7.8.71
25. Barac A, Holik M, Ćurić K, Stojkov M (2026) Hybrid renewable systems integrating hydrogen, battery storage and smart market platforms for decarbonized energy futures. Energies 19:331. https://doi.org/10.3390/en19020331
26. Ramdan M, Hakim AL, Ramdani S, Wicaksana FA (2025) IoT integration for renewable energy storage: a systematic literature approach. Eng Proc 107:101. https://doi.org/10.3390/engproc2025107101
27. Borges Schiavon H, Peretti Correa D, Lissa P, Keane M (2025) Advances and challenges in P2P energy trading: a review of renewable integration and decarbonisation in Ireland. J Phys Conf Ser 3140:032008. https://doi.org/10.1088/1742-6596/3140/3/032008
28. Al Jasem AMS, De Clark T, Shrestha AK (2025) Toward decentralized intelligence: a systematic literature review of Blockchain-enabled AI systems. Information 16:765. https://doi.org/10.3390/info16090765
29. Muiz Tairu A, Oluwatosin Junior A, Akindele TE, Akeem S (2025) The role of Blockchain-based smart contracts in enhancing financial transparency and efficiency in the emerging market. IJRIAS 10(9). https://doi.org/10.51584/IJRIAS.2025.100900052
30. Santos L, Gomes A, Rupino P (2025) Energy trading using blockchain: smart contracts functionalities—a systematic review. Energ Strat Rev 61:101825. ISSN 2211-467X. https://doi.org/10.1016/j.esr.2025.101825
31. Amratanshu JA, Jaiswal A (2025) Blockchain in energy: overcoming barriers and harnessing opportunities for a decentralized future. World J Adv Eng Technol Sci 15(01):1300–1307. https://doi.org/10.30574/wjaets.2025.15.1.0333
32. Uddin M, Obaidat M, Manickam S, SUA L, Dandoush A, Ullah H, Ullah SS (2024) Exploring the convergence of Metaverse, Blockchain, and AI: a comprehensive survey of enabling technologies, applications, challenges, and future directions. WIREs Data Mining Knowl Discov 14(6):e1556. https://doi.org/10.1002/widm.1556
33. Weking J, Desouza KC, Fielt E, Kowalkiewicz M (2023) Metaverse-enabled entrepreneurship. J Bus Ventur Insights 19:e00375. ISSN 2352-6734. https://doi.org/10.1016/j.jbvi.2023.e00375
34. Amin U, Kim D, Ahmed FN, Ahmad G, Hossain MJ (2025) Digital twins for smart asset management in the energy industry: state-of-the-art. Expert Syst Appl 289:128358. ISSN 0957-4174. https://doi.org/10.1016/j.eswa.2025.128358
35. Kumba K, Ra N, Goswami A, Gundu V (2026) Beyond the grid: a decade of intelligence, interoperability, and innovation in smart energy networks. Int J Sustain Energy 45(1). https://doi.org/10.1080/14786451.2026.2622726

36. Ishfaq H, Kanwal S, Anwar S, Abdussalam M, Amin W (2025) Enhancing smart grid security and efficiency: AI, energy routing, and T&D innovations (a review). Energies 18:4747. https://doi.org/10.3390/en18174747
37. Alhasnawi BN, Hashim HK, Zanker M, Bureš V (2025) The rising, applications, challenges, and future prospects of energy in smart grids and smart cities systems. Energy Convers Manage X 27:101162. ISSN 2590-1745. https://doi.org/10.1016/j.ecmx.2025.101162
38. Udo WS, Kwakye JM, Ekechukwu DE, Ogundipe OB (2023) Predictive analytics for enhancing solar energy forecasting and grid integration. Eng Sci Technol J 4(6). https://doi.org/10.51594/estj.v4i6.1394
39. Javed H, Eid F, El-Sappagh S et al (2025) Sustainable energy management in the AI era: a comprehensive analysis of ML and DL approaches. Computing 107:132. https://doi.org/10.1007/s00607-025-01485-0
40. Campana P, Censi R, Ruggieri R, Amendola C (2025) Smart grids and sustainability: the impact of digital technologies on the energy transition. Energies 18:2149. https://doi.org/10.3390/en18092149
41. Bhutta MS, Li Y, Abubakar M et al (2024) Optimizing solar power efficiency in smart grids using hybrid machine learning models for accurate energy generation prediction. Sci Rep 14:17101. https://doi.org/10.1038/s41598-024-68030-5
42. Navarro MC, Fernández-Guillamón A, Sarasua JI, Gómez LS, Gil-García IC, Martínez-Lucas G (2025) Energy technologies for hybrid renewable energy systems: a study of spatially viable sites for hybrid power plants in Spain. Front Energy Res 13:1628824. https://doi.org/10.3389/fenrg.2025.1628824
43. Hossain MdS, Abido MA (2025) Stability prediction in sustainable energy systems using machine learning models. Franklin Open 11:100260. ISSN 2773-1863. https://doi.org/10.1016/j.fraope.2025.100260
44. Lifelo Z, Ding J, Ning H, Qurat-Ul A, Dhelim S (2024) Artificial intelligence-enabled Metaverse for sustainable smart cities: technologies, applications, challenges, and future directions. Electronics 13:4874. https://doi.org/10.3390/electronics13244874
45. Maier F, Weinberger M (2024) Metaverse meets smart cities—applications, benefits, and challenges. Fut Internet 16:126. https://doi.org/10.3390/fi16040126
46. Bandeiras F, Gomes Á, Gomes M, Coelho P (2023) Exploring energy trading markets in smart grid and microgrid systems and their implications for sustainability in smart cities. Energies 16:801. https://doi.org/10.3390/en16020801
47. Chen Z, Gan W, Wu J, Lin H, Chen CM (2024) Metaverse for smart cities: a survey. Internet Things Cyber-Phys Syst 4:203–216. ISSN 2667-3452. https://doi.org/10.1016/j.iotcps.2023.12.002
48. Antal C, Cioara T, Antal M, Mihailescu V, Mitrea D, Anghel I, Salomie I, Raveduto G, Bertoncini M, Croce V, Bragatto T, Carere F, Bellesini F (2021) Blockchain based decentralized local energy flexibility market. Energy Rep 7:5269–5288. ISSN 2352-4847. https://doi.org/10.1016/j.egyr.2021.08.118
49. Zahid H, Zulfiqar A, Adnan M, Iqbal MS, Shah A, Abbasi U, Mohamed SEG (2025) Transforming nano grids to smart grid 3.0: AI, digital twins, blockchain, and the metaverse revolutionizing the energy ecosystem. Results Eng 27:105850. ISSN 2590-1230. https://doi.org/10.1016/j.rineng.2025.105850
50. Florentin-Robert D, Toma L, Picioroaga I (2026) The Smart City from the energy perspective. Preprints 2026011925. https://doi.org/10.20944/preprints202601.1925.v1
51. Sharifi A, Amirzadeh M, Khavarian-Garmsir AR (2025) The metaverse as a future form of smart cities: a systematic literature review of co-benefits and trade-offs for sustainable development goals. Cities 161:105879. ISSN 0264-2751. https://doi.org/10.1016/j.cities.2025.105879
52. Rangarajan SS et al (2023) DC microgrids: a propitious smart grid paradigm for smart cities. Smart Cities 6(4):1690–1718. https://doi.org/10.3390/smartcities6040079

53. Yaqoob I, Salah K, Jayaraman R, Omar M (2023) Metaverse applications in smart cities: enabling technologies, opportunities, challenges, and future directions. Internet of Things 23:100884. ISSN 2542-6605. https://doi.org/10.1016/j.iot.2023.100884
54. Bhavana GB, Anand R, Ramprabhakar J et al (2024) Applications of blockchain technology in peer-to-peer energy markets and green hydrogen supply chains: a topical review. Sci Rep 14:21954. https://doi.org/10.1038/s41598-024-72642-2

Chapter 3
Metaverse Energy Sector Factories

Abstract The energy or renewable energy virtual factories or plants (EREVFs or Ps) have been recently proposed as means to serve an optimized, ecological and cost-effective energy sector management, globally. However, related research, innovation and enterprise data (RIED) or actionable industrial site use cases seem to be rather rare in range and depth. In a similar vein, there is scarce evidence that aligns the projected Metaverse-enabling technologies, application programming interfaces (APIs) and human-intelligence interaction (HII) [e.g., Digital Twin (DT), Internet of Things (IoT), Blockchain (B), Devices (D), Edge and Cloud Computing (ECC), Artificial Intelligence (AI), Machine Learning (ML), Virtual Reality (VR), Augmented Reality (AR), Extended Reality (XR)] with the forecasted smart energy (renewable energy) grid and the virtual factories or plants, in particular. In that respect, the current chapter aims to outline our future, operationalized, "Energyverse" or "Renewenergyverse" ecosystem. This higher-level, distinct and open-eco, smart innovation infrastructure might calibrate the multi-energy (renewable energy) distributed resource generation, storage and transmission through multiple channels or stakeholders. For example, wind turbines, hydro plants, solar panels, electric vehicles, utilities, investors, founders, consumers, communities, government, offices, residences and regulatory bodies, among others. It is connected to the intelligent grid on a control management system, monitoring, smart intelligence and marketplace components. This inclusive network of Metaverse virtual factories or plants is envisioned to support an optimized, secure, affordable and stable energy (renewable energy) cycled grid ecosystem. The current and future research and innovation need to be developed to offer deeper insights on the benefits of such immersive energy (renewable energy) optimized intelligence, controlling for potential concerns over computational intelligence, ethics, safety, affordability and sustainability.

Keywords "Renewenergyverse" · "Energyverse" · Virtual energy factories or plants · Distributed renewable energy · Electric vehicles · Energy optimization · Metaverse-enabling information and communication technologies · Security · Cost · Open innovation

E. G. Makri, *Metaverse and Smart Grid in Energy Industry: An Open Innovation Challenge on Trends and Applications*, Studies in Smart Technologies,
https://doi.org/10.1007/978-3-032-29666-5_3

3.1 Introduction

In recent years, escalating climate disruption, unstable energy grid, greater energy demands and less affordable energy consumption, have fueled international research and innovation (RI) on alternative, less costly and more ecological energy management. The distributed renewable energy resources, the Metaverse-enabling technologies and the virtual factories or plants (VFs or Ps), constitute tentative pillars of an envisioned, next frontiers, open innovation (i.e., outflow to inflow), climate-friendly and sustainable energy research, innovation and enterprise (RIE). One could reflect upon the potential benefits that this projected, holistic, "Energyverse" or "Renewenergyverse" might have in serving an optimized, environment-friendly and of advanced computational intelligence, affordable energy system. Yet, relevant industrial case studies and empirical evidence tend to be rather either scarce, limited or divided in scope and depth. For instance, such data are either projected on specific Metaverse-driven technologies in the integrated energy sector (e.g., DT, IoT, B, AI, ML, etc.) or the stability of the energy grid due to distributed renewable energy (DRE) or the operation of the EREVFs or Ps. Nevertheless, the above inclusive ecosystem is envisioned to operate through a distant energy (or renewable energy) supplier (e.g., utilities, marketplace) control system (ESCS or RESCS) connected to the VF or VP with shared energy (or renewable energy) generation, storage and distribution. This forecasted energy (or renewable energy) production, maintenance and transmission cycled process is projected to be intertwined across multimodal pathways. Indicative examples of such paths could include, for instance, various renewable energy resources (e.g., wind turbines, hydro plants, solar panels, electric vehicle stations, etc.) and (or) different consumer routes (e.g., residences, offices, factories, government, agencies, communities, etc.). The following sections describe the forecasted "Energyverse" (EVerse) or "Renewenergyverse" (REVerse) simulation as an operationalized, distinct and integrated ecosystem with EREVFs or Ps that could offer foundational advantages to an intelligent energy domain management, if calibrated along with corresponding technical, monitoring, regulatory and cost-effective principles.

3.2 Renewable Energy Factories or Plants

We envisage that the challenges (or concerns) associated with garnering and handling the energy produced from distributed renewable energy resources (DRERs) (e.g., wind turbines, solar panels, hydro plants, electric vehicle recharging hubs, etc.) are well reported [1]. To address those, we argue that the energy (or renewable energy) virtual factories or plants are proposed as next generation innovations. Such transformative or disruptive technological systems might feed the optimization of energy (or renewable energy) running ecosystem. This could be achieved through the aggregation of the existed renewable energy resource types (i.e., as indicated above), as

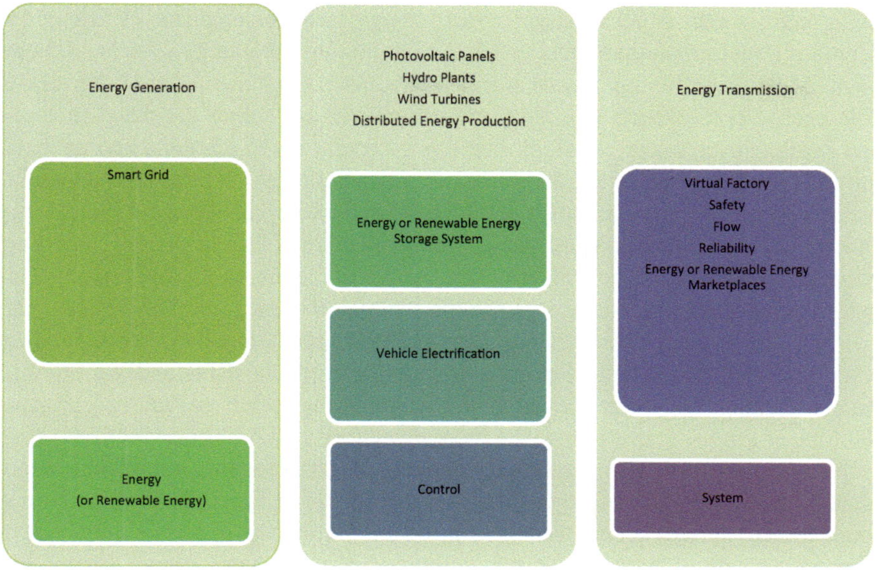

Fig. 3.1 Energy (or renewable energy) virtual factory or plant

focal point of reference. In that respect, they could form a rather calibrated, safer and cost-manageable, ecological energy infrastructure [2]. In that sense, there could be a balance between the functional expenses adhered to the energy (or renewable energy) collection from suppliers and the potential shortage conditions met during energy (or renewable energy) storage (e.g., [3]). Our own tentative, envisioned energy (or renewable energy) virtual factory or plant is depicted in the plain-sketched Fig. 3.1.

The aforementioned energy (or renewable energy) virtual factory or plant comprises of three sub-sections. The first constitutes the energy or distributed renewable energy resources (EDRERs) which can be integrated (or interrelated) within a distinct ecosystem on its own. The second forms the energy (or renewable energy) storage infrastructure (ERESI) that might be adaptable to fluctuations or demands. The third module is composed of the energy (or renewable energy) control (monitoring) system of the virtual factory or plant. This system represents the cross-over (or in-between) information and communication flow across the EDRERs and ERESI. The end result of the virtual factory or plant is projected to offer a more optimized, secure, affordable and controlled usage management of the overall energy (or renewable energy) ecosystem. This network can be envisaged to feed into smart grid, extend to markets and (or) be further calibrated with AI and ML solutions that can precisely forecast, plan and arrange energy and (or) renewable energy needs. Nevertheless, it is an envisioned, extended infrastructure that might be explored in different, applicable industrial energy cases, utilities, regions, countries or communities.

In recent years, small or larger virtual energy factories or plants (VEFs or Ps) denote a transformational development in optimizing the energy system management. Most recently, this advancement is gradually implemented into the renewable energy optimization [4], although still slowly and rather restricted in scope and depth. Nevertheless, we think that this process aims to integrate and calibrate the eco-friendly (renewable), distributed energy resources (RDERs) (and demands) inclusively, serve the optimization of the networked energy management (i.e., energy generation, storage and transmission), the energy marketplace (e.g., energy trading, costs, etc.) and improve the scalability and control of the smart (energy) grid [5]. This optimized procedure appears to be feasible through the integration of related smart energy technological advancements, decentralized foundational and consumer-side resources, services and tools, supported by monitoring mechanisms, and real-timeline data processing. The limited research on renewable energy virtual factories or plants (REVFs or Ps) [6] seems to reveal a more diversified and multi-perspective system architecture. Nevertheless, we operationalize a generic model of such factories (or plants) "mirroring" our own visual conceptual model in Fig. 3.2. The intertwined foundational (infrastructure) and platform tiers "echo" the corresponding renewable energy virtual factory or plant monitoring system management, virtual API interface, intelligent data anthology, diffusion, analysis and visual user interactive experience. At a higher and integrated level of "framing", the factory or plant runs on the distributed ecological energy resources provided by the connected suppliers (i.e., utilities, investors, founders, etc.) and aligns with the matching market trading, consumer houses, regional companies, communities and regulatory associations, as sketched in Fig. 3.2 that follows.

3.3 Renewable Energy Factories or Plants on Metaverse

We claim that recent though sparse work on energy generation technology seems to propose Metaverse as an umbrella (inclusive) term that envisions its implementation as the projected computational modelling platform that could empower or fuse the future energy systems, globally (e.g., [7, 8]). Re-configuring our operationalization, we feel that Metaverse can be envisaged as the next frontiers intelligent ecosystem. Thus, it is critical not only to integrate distributed energy, but also to integrate dispersed renewable energy through virtual factories or plants for a more optimized, affordable, controllable and protected energy industry infrastructure, embedded into Metaverse (e.g., [9, 10]). This ecosystem might embrace the energy or renewable energy infrastructure (encompassing the VE or REVFs or Ps), in terms of the human-machine (intelligent) user interface or experience (UI or UX) aspect, at a universal scale. Figure 3.3 outlines this explorative, projected conceptualization, as next.

Within our aforementioned infrastructure, twofold information and communication diffusion enables scalable interaction between the physical (real-world) and virtual energy (or renewable energy), digital innovation systems. The intelligent

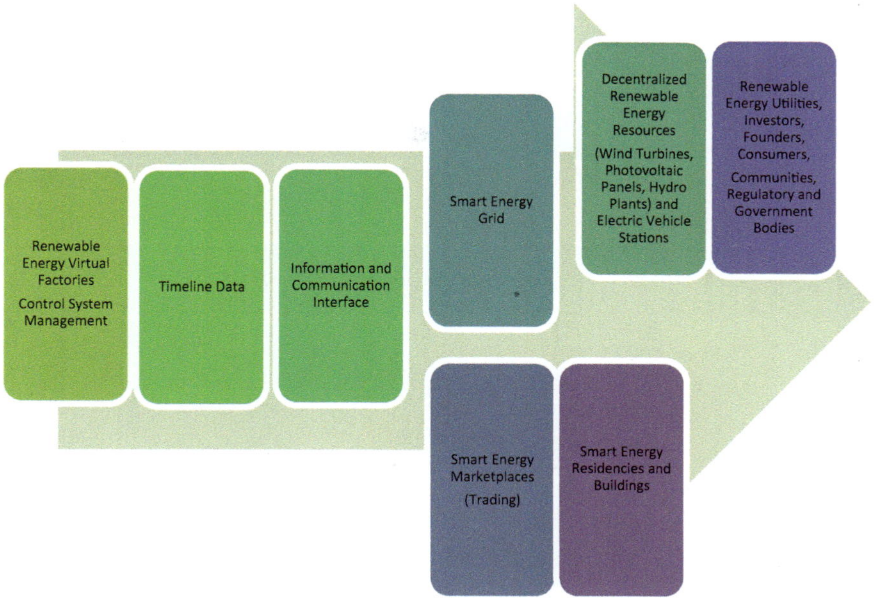

Fig. 3.2 An operationalized model of renewable energy virtual factories or plants

(or smart) analyses integrate and examine data from both the physical and virtual spheres to generate related outputs or decision-making. The human-intelligent interface supports energy (or renewable energy) consumers (or suppliers, communities, stakeholders) to get access to and exchange smart (intelligent) information (or data) relevant to both physical and virtual energy (or renewable energy) systems.

The Metaverse interface environment tier includes the real-world, the virtual energy (or renewable energy) systems and the human-intelligent interaction domains. In the architecture tier, the physical energy (or renewable energy) system feature shapes the foundational frame for the virtual energy (or renewable energy) system to assemble. The latter acts as a digital replica in the energy Metaverse or "Renewenergyverse" ecosystem, adaptable to timeline updates from the physical energy (or renewable energy) system. The core information and communication technologies (ICT) that serve the implementation of the virtual energy (or renewable energy) systems and the human-intelligent interaction operation encompass the DT and the IoT. The latter leverage their associated inclusive devices (physical equipment), edge (linked network), cloud (hub-based, service system), visual communication, sensing platforms and databases solutions. In that effect, they can transmit, analyse, monitor and handle captured data and potential energy fluctuation discrepancies, thus, serving energy sustainability. AI and ML supporting technology with their sub-technologies can further address energy (or renewable energy) projected production demands, monitoring, run and smart grid error diagnosis and intervention.

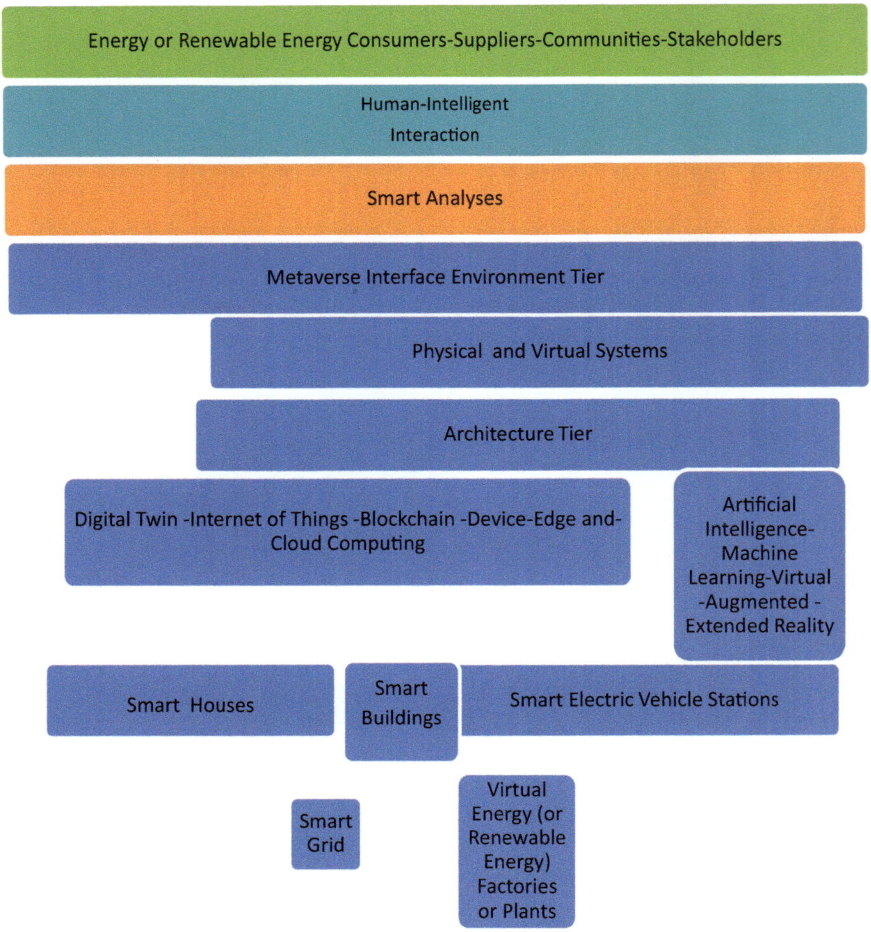

Fig. 3.3 A "human-intelligent" energy Metaverse or "Renewenergyverse" ecosystem including virtual energy (or renewable energy) factories or plants

The smart analyses component on the Metaverse interface environment tier supports the intelligent energy (or renewable energy) system to run, using computational power (CP) and AI. The human-intelligent interaction domain facilitates the information and communication flow between the energy Metaverse or "Renewenergyverse" ecosystem and its clients (i.e., consumers, utilities, investors, founders, communities, regulatory and government agencies, etc.), with XR supporting its application. In the architecture tier, the association between the physical system, the virtual system and the VR exchange "mirrors" the interaction across AR and VR, enabling the twofold visual and sensing information and communication flow from the virtual energy to the real-world systems, fostered by the VR mode.

The virtual system appears to monitor the physical system through analogous DT visual, optimized, data-based flow and screening, providing with appropriate virtual energy (or renewable energy) factory or plant affordable and reliable development settings. In addition, XR digitalization tends to support functional updates for energy (or eco-friendly energy) devices in the factory or plant, allowing them to garner visualized, simulated, space and time data to be processed. The aforementioned smart (or intelligent) analyses further served by the human-intelligent exchange seem to constitute or being applied as distinct parts within physical energy (or renewable energy) ecosystems.

Envisioning the above-operationalized human-intelligent energy Metaverse or "Renewenergyverse" virtual factory or plant ecosystem, it might be envisaged to include certain challenges. Inter alia, first, its energy (or renewable energy) data nature, consistency and control. Second is the scalability of mind mapping ecosystem simulations. Third, the updates on regulatory frameworks. Fourth is the advanced security screening and real-time interventions. Fifth is the improved computational modelling that serves low latency and affordable cost. Successfully addressing the above optimization-driven concerns, a controllable, operational and safer energy (or renewable energy) virtual factory or plant management system might be envisioned.

Notwithstanding the above, Roozbehani et al. [11] "echo" the VEFs or Ps as emerging energy production systems feasible to address the energy management challenges associated with the dispersed eco-friendly (i.e., renewable) energy resource utilization across the energy sector infrastructure, globally. They "mirror" the relevant factories as powerful energy ecosystems still in infancy that can include core and subsidiary-led resources of dispersed and even renewable energy marketplace production, maintenance and distribution. In that sense, we envisage that the VEFs or Ps can organize, control and optimize the function of their constituent energy resources, energy budget, carbon emissions and monetized energy resource(s) income. These factories tend to be visually conceptualized in diverse versions, build upon varied energy and renewable resource parts or features across the smart energy sector grid. Our own operationalized, tentative visual conceptual model of them is envisioned in the following Fig. 3.4 outlined, as next.

Further reflecting upon and analysing the above authors' rationale, we think that the optimized configuration and function of the VEFs or Ps seems to gradually rest upon the technological, ecological and financial motives (and objectives) of the connected energy suppliers, clients and energy governance agencies. Such drives and goals can vary between reducing the overall energy expenses on the ecosystem for the customers, to limiting energy waste, error handling, carbon discharges and improving the total energy infrastructure operation. However, "echoing" the above authors' insightful accounts, the VEFs or Ps seem to be related to tentative concerns or risks connected to the overall smart energy grid's security and steadiness (p. 5–6 above). That is, the ambiguity that surrounds the eco-friendly energy resource(s) timeline life-cycles, the uncertainty that defines the energy value marketplaces and the insecurity that spans across the configuration of the energy workload needs (p. 1 above). Nevertheless, the VEFs or Ps can support the actual timeline, direct and indirect, external to internal (i.e., open innovation), blockchain-based, distributed,

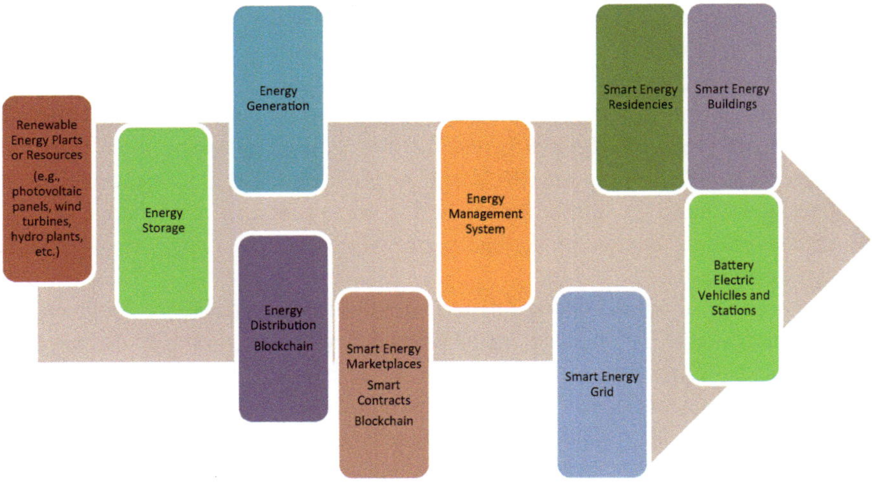

Fig. 3.4 Virtual energy factories or plants operationalized visual conceptual model

smart (i.e., intelligent, efficient, information and communication-balanced) and safer energy exchange between energy suppliers, clients, operators and regulatory bodies. In this respect, energy producers can sell their energy assets. Energy consumers can also buy and sell, for example, their extra renewable and (or) electrical vehicle energy assets (i.e., properties) in the energy marketplaces, without the involvement of other agents, concurrently. This process is managed through smart contracting. The latter tends to secure the energy costs configuration being restricted after energy distribution error handling. As a result, energy utility organizations can calibrate their functional, balanced energy generation, storage and distribution expenses by leveraging on environmentally–friendly energy resources. Nevertheless, the aforementioned authors "echo" the utilization of supporting networks or services on the VEFs or Ps in varied timeline energy demands life-cycles. In our reflective insight, this can be performed, for instance, by using a range of different, available renewable resources to safeguard efficient and sustainable system operation and monitoring [12]. In their concluding reflection that "mirrors" indicative, next frontiers research and innovation needed on VEFs or Ps, the first authors envision the exploration of convolutional neural network models (CNNMs) and additional control mechanisms to enable the sustainable operation and monitoring of such energy factories or plants (p. 12, above).

By and large, we think that the VEFs or Ps are depicted as future, forecasted ecosystems of energy (renewable energy) sector and assets, blockchain-core and information and communication-driven ones. They are envisioned to operate within manufacturing location(s) aimed to reduce (or optimize) energy expenditure, greenhouse discharges and overall energy utilization by producing, maintaining, distributing and trading calibrated, renewable energy resource units in-between and across the smart energy grid and the connected energy suppliers [13]. "Echoing" our

Fig. 3.5 Virtual energy factories or plants on industrialized energy sector metaverse

above reflective arguments, our own conceptualized future VEFs or Ps embedded into Metaverse-industrialized smart (intelligent) energy domains are sketched in Fig. 3.5 that comes next. These factories or plans are envisaged as parts of a life-cycled, integrated "Energyverse" network, enabled by Metaverse technologies, run on a control system, connected to the intelligent grid, fueled by and aligned with multiple renewable energy resources, linked stakeholders and trading marketplaces.

Analysing the authors' [14] operationalization of enterprise renewable energy Metaverse (p. 4459), among other scholars, we argue that the envisioned Metaverse infrastructure can provide energy stakeholders (and communities) with an inclusive, digital innovation ecosystem into the smart energy grid. Such eco-platform can serve the sustainable design, development, assessment, maintenance and distribution of powerful, eco-friendly and decentralized energy ecosystems, globally. However, such an "Energyverse" or "Renewenergyverse" (i.e., the energy or renewable energy sector industrial or enterprise Metaverse) tends to require advanced and multimodal energy solutions. Such systems seem to demand complicated energy and renewable energy resources, services and support applications, which are currently very limited in exploration, scale or regulation.

Reflecting on the above, we claim that a tentative account of such rare, Metaverse renewable energy solutions is projected by [5, 15, 16] and [17]. They envision the forecasted design and deployment of VEFs or RVEFs or Ps (e.g., photovoltaic). In reflecting upon their VREFs or Ps use cases developed ecosystems, we note the following insights, as next. First, an array of diverse Metaverse technologies, front- and back-end application, information and communication, gaming and climate programming interfaces and simulations is used to serve the computational architecture of their infrastructure. Second, their fully-fledged intelligent power design is supported by visualization tools, real-timeline data, error screening and cybersafety tools. Expanding our reflection upon the above authors' VREFs or Ps simulation(s) on Metaverse, we outline our own "Renewenergyverse" VREF or P infrastructure in the following Fig. 3.6. The latter "mirrors" the umbrella (overarching) ecosystem that serves the eco-friendly, "smart" and "optimized" VREF or P projection. This projection is intertwined with cross-over software and hardware APIs and tools, integrated into the Smart Energy Grid eco-platform and utilized by multimodal stakeholders (e.g., investors or founders, clients, utilities, communities, regulatory and government agencies, etc.). Our own further envisioned architecture of this "intelligent" projected ecosystem that depicts a "Renewenergyverse" front-and-back-end VF or P modelling is sketched in Fig. 3.7 that follows next. This "optimized-driven" and "smart" projection is served by foundational and UI (or UX) Metaverse-led technologies, tools and components that aim to support the infrastructure's scalability, security and client-interaction streamlined experience, efficiently.

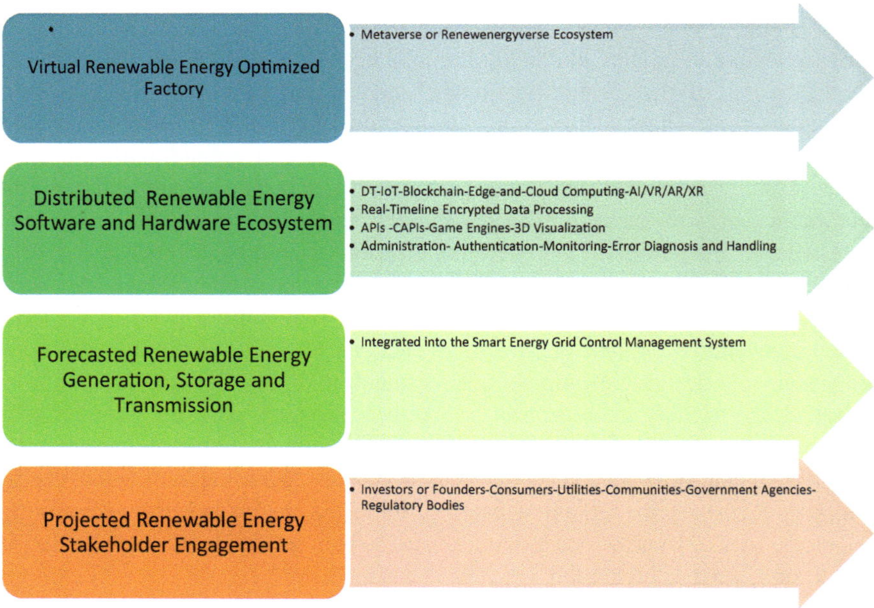

Fig. 3.6 An envisioned "Renewenergyverse" optimized virtual factory or plant

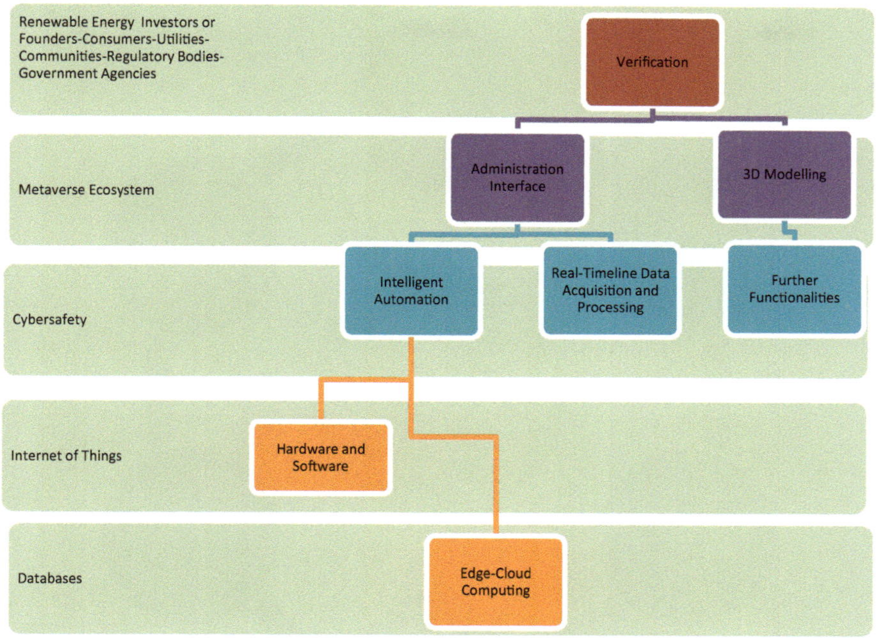

Fig. 3.7 A reflection of the "Renewenergyverse" model structure

3.4 Conclusion

Our envisioned, eco-friendly EVF or P might be projected to run on a distributed, renewable energy ecosystem on visualized Metaverse. This higher-order Metaverse infrastructure could "mirror" an open innovation (OI) framework. That is, the integration (or fusion) of external to internal (i.e., outbound to inbound, outflow to inflow) renewable energy resources or supplier networks and associated stakeholders into the Metaverse intelligent grid and trading marketplaces. The operation and control of this 3D visual ecosystem could be made feasible, optimized, reliable and affordable through the application of the prior documented and aligned Metaverse and ICTs-enabling immersive APIs and human-intelligent interface(s). Following the aforementioned, current scarce-related data and studies should be further empowered to include Metaverse-led VFs or Ps renewable energy simulations. This would not only serve the provision of progressive data and attributes on the actionable, optimized, reliable and affordable generation, operations, storage, distribution and usage management of the "Energyverse", renewable VFs or Ps but it would also support a deeper exploration, RIE engagement and calibration of any associated challenges. Such concerns might entail, for example, issues of computational modelling scalability, security, costs, renewable energy resource stable generation, storage, transmission, suppliers and marketplace trading integration, regulatory standards and monitoring, inter alia. In a nutshell, resting on previous argumentation, our concluding

Fig. 3.8 Our open innovation-driven energy or "Renewenergyverse" virtual factory or plant

reflective insight of the future "Energyverse" or "Renewenergyverse", OI VF or P is sketched in the following visual mind map (Fig. 3.8).

References

1. Çaloğlu Büyükselçuk E, Turan H (2026) evaluation of barriers to the integration of renewable energy technologies into industries in Türkiye. PRO 2026 14:307. https://doi.org/10.3390/pr14020307
2. Hasan M, Zarin NA, Rezwan Ahmed M, Farrok O (2026) Global pathways for hybrid renewable energy systems: challenges, solutions, policy, and regulatory frameworks. Energy Convers Manag X 29:101534. ISSN 2590-1745. https://doi.org/10.1016/j.ecmx.2026.101534
3. Borisoot K, Liemthong R, Srithapon C, Chatthaworn R (2023) Optimal energy management for virtual power plant considering operation and degradation costs of energy storage system and generators. Energies 16:2862. https://doi.org/10.3390/en16062862
4. Zhang P, Zheng Y (2026) Economical management of virtual power plant source-network-load-storage in the context of electricity-carbon market. PLoS One 21(1):e0338321. https://doi.org/10.1371/journal.pone.0338321

5. Abdelkader S, Amissah J, Abdel-Rahim O (2024) Virtual power plants: an in-depth analysis of their advancements and importance as crucial players in modern power systems. Energ Sustain Soc 14:52. https://doi.org/10.1186/s13705-024-00483-y
6. Skoczkowski T, Bielecki S, Wołowicz M, Węglarz A (2025) Redefining energy management for carbon-neutral supply chains in energy-intensive industries: an EU perspective. Energies 18:3932. https://doi.org/10.3390/en18153932
7. Huang C, Bu S, Chen Q, Lee HH (2023) Meta-power: digitalized power systems driven by Metaverse. J Power Gener Technol 1(2):20–49. https://doi.org/10.61369/JPGT.7355
8. Shimpo F, Sokołowski MM, Mikusek P, Chawla Y, Figueroa Jr RB (2025) Cybernetic avatars for energy transformation: regulating the digital revolution for enhancing sustainable attitudes of energy consumers. Digital Policy Regul Govern 28(1):35-51. ISSN 2398-5038. https://doi.org/10.1108/DPRG-12-2024-0320
9. Han Y, He X, Zhang H et al (2024) Metaverse framework designing for energy scheduling in energy internet of things considering emergence. Digital Twin 3:6. https://doi.org/10.12688/digitaltwin.17873.2
10. Naval N, Yusta JM (2021) Virtual power plant models and electricity markets—a review. Renew Sust Energ Rev 149:111393. https://doi.org/10.1016/j.rser.2021.111393
11. Roozbehani MM, Heydarian-Forushani E, Hasanzadeh S, Elghali SB (2022) Virtual power plant operational strategies: models, markets, optimization, challenges, and opportunities. Sustainability 14:12486. https://doi.org/10.3390/su141912486
12. Darvishi M, Tahmasebi M, Shokouhmand E, Rasupuleti J, Bokoro P, Raafat JS (2023) Optimal operation of sustainable virtual power plant considering the amount of emission in the presence of renewable energy sources and demand response. Sustainability 15:11012. https://doi.org/10.3390/su151411012
13. Urbano EM, Martínez Viol V (2020) Energy infrastructure of the factory as a virtual power plant: smart energy management. In: New trends in the use of artificial intelligence for the Industry 4.0. IntechOpen, pp 1–22. https://doi.org/10.5772/intechopen.88861
14. Raman R, Kautish P, Siddiqui A, Siddiqui M, Nedungadi P (2025) The role of metaverse technologies in energy systems towards sustainable development goals. Energy Rep 13:4459–4476. https://doi.org/10.1016/j.egyr.2025.04.012
15. Yalçin T, Paradell Solà P, Stefanidou-Voziki P, Domínguez-García JL, Demirdelen T (2023) Exploiting digitalization of solar PV plants using machine learning: digital twin concept for operation. Energies 16:5044. https://doi.org/10.3390/en16135044
16. Massel L, Shchukin N, Cybikov A (2021) Digital twin development of a solar power plant. E3S Web Conf 289:03002. https://doi.org/10.1051/e3sconf/202128903002
17. Liu X, Gao C (2025) Review and prospects of artificial intelligence technology in virtual power plants. Energies 18:3325. https://doi.org/10.3390/en18133325

Chapter 4
Metaverse Energy Marketplaces-
"EnergyMarketverse"

Abstract In the present global smart technology intelligence, sustainable and eco-friendly research, innovation, enterprise and outreach (RIEO), the paradigm of diverse nature, envisioned, decentralized Blockchain (B)-driven smart energy trading platforms that could be applied on Metaverse, tends to be rare and limited in scale. Thus, the current chapter brings forward this extensive gap in a fourfold alignment, as next. First, by exploring and reviewing the scarce, evidence-based research, innovation and impact (RII) across the multiple, monetized, non-fungible tokens (NFTs), smart contract (SC) and sensing-led, B-diverse energy trade product or service consumer and supplier networks. Such systems are envisaged to be implemented on the envisioned future "EnergyMarketverse" infrastructure. Second, by highlighting the dynamics of the eco-sustainable nature and effect(s) of these future, multimodal, NFTs-capitalized, SC, sensing and B-distinct "EnergyMarketverse" (product or service) trading (customer or producer) networks. Third, by operationalizing these next frontiers multifold, valuated (NFTs-based), SC-and-sensing, B-divergent "EnergyMarketverse" (asset or service) trading (client or supplier) ecosystems, as open innovation ones. Forth, by reviewing the benefits and challenges that these next generation, multilevel, NFTs-investment, SC and sensing-driven, B-cross-over "EnergyMarketverse" (product or service) trading (customer or producer) infrastructures might entail. In specific, reflecting upon the future "EnergyMarketverse", multimodal, technical, eco-sustainability and regulatory framework operational concerns, inter alia (i.e., computational architecture, accessibility, scalability, transparency, security, reliability, maintenance, interoperability, monitoring, cost-efficiency, utilization of renewable energy resources, regulation and authentication endorsement).

Keywords Metaverse · "EnergyMarketverse" · Blockchain · Energy trading · Non-fungible tokens · Smart contracts · Sensing · Renewable resources · Sustainability

E. G. Makri, *Metaverse and Smart Grid in Energy Industry: An Open Innovation Challenge on Trends and Applications*, Studies in Smart Technologies,
https://doi.org/10.1007/978-3-032-29666-5_4

4.1 Introduction

Asif and Hassan [1] in their overview of the Internet of Things (IoT) and Metaverse embeddedness outline the contribution of the Blockchain (B) immersive infrastructure enabling the co-creation (i.e., open innovation) of virtual financial systems and marketplaces on Metaverse. Across this new era of virtual trading service(s) and user interface(s), customers and (or) providers might be able to sell, buy, import and (or) export diverse energy virtual goods, assets, entities and (or) services. Energy trading exchange and associated economic transaction(s) may emerge as outcomes of this energy supplier-end-user Metaverse interaction. In the current state, the accelerating growth of research and co-development of energy Metaverse-led platforms tends to be still scarce in scope and range. Therefore, there seems to be an expanded gap that needs to be covered to foster Metaverse-based energy trade and monetization-led circular economy research, innovation and practice in the upcoming years.

Along the above vein of Metaverse-driven envisioned energy immersive ecosystem, Nleya and Velempini [2] suggest that the future energy Metaverse is expected to integrate the following energy widespread sub-networks, as next. First, the energy customer(s) Metaverse reflecting recreation and wholesaling. Second, the energy industrial Metaverse (p. 2 above) that integrates all aspects of energy product(s) or service(s) development, assessment and delivery. Third, the energy enterprise Metaverse (as above) that enables the energy workforce networked professional interaction(s). Extending the authors' aforementioned envisioning of energy detached Metaverse industries and research and development (R&D) indicating use cases of environmentally-sustainable industrial Metaverse implementation (applications, as above p. 25), we may consider the following reflections, as next. First, we might envisage an open-innovation-driven energy Metaverse infrastructure as one, incorporating the above 3 energy-related properties as an interconnected, single open-innovation-led entity [i.e., external-to-internal flow of resources (agents, assets, capabilities) as conveyors of innovation, for instance [3–5]. Second, this "alternative" energy Metaverse of interrelated technologies, operations and R&D might be "projected" across various real-life use cases by relevant businesses and customers [i.e., energy service(s) and (or) consumer(s)]. This "projection" might explore and (or) disclose potential constructive impact for energy expenditure (trading, e-waste recycling, etc.) following-up on prior positive and environmentally-sustainable evidence-based industrial Metaverse effects (see above). Third, the "projected" energy Metaverse could create more opportunities for businesses and customers to adopt more environment-friendly strategies. These policies might be based, among others, on renewable resources and (or) sensing and blockchain technologies to build energy "fungible digital assets" (p. 23 above), "mirroring" eco-friendly industrial Metaverse outcomes (as above). Figure 4.1 sketches an outline of our tentative, overall, energy Metaverse ecosystem operationalized as open innovation one, as next.

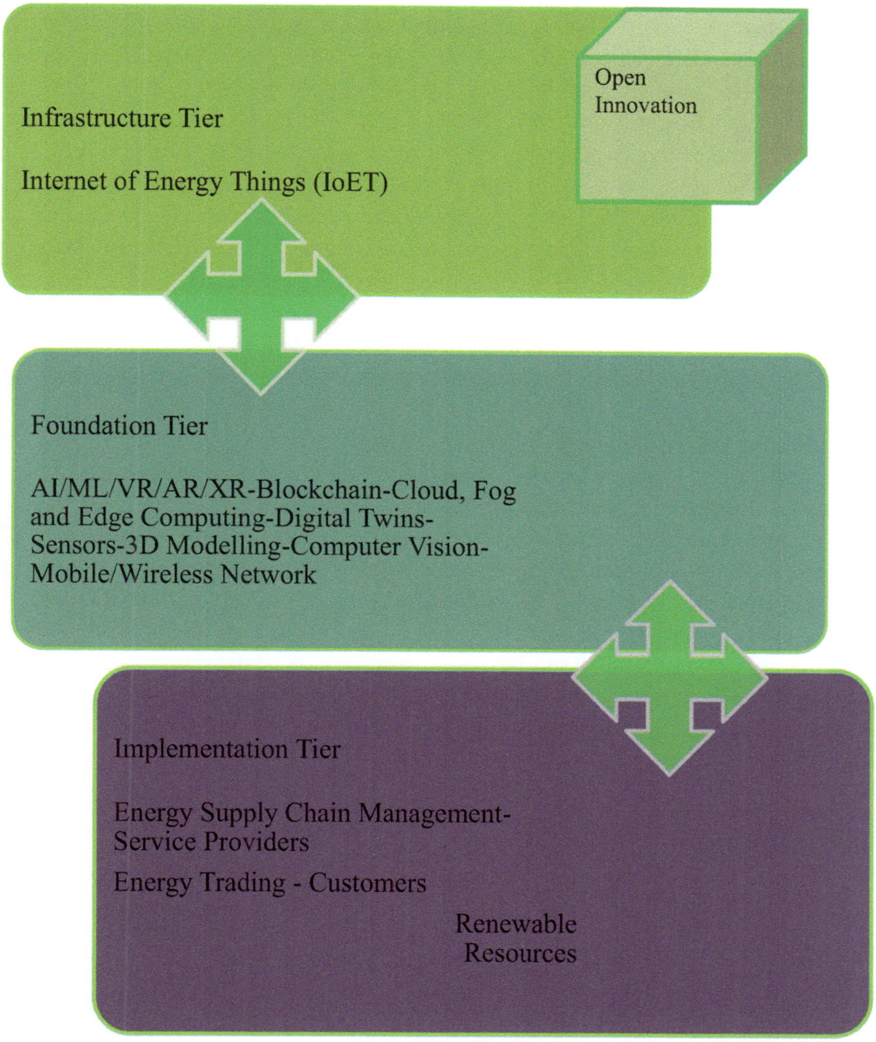

Fig. 4.1 Outline of the energy open innovation-led Metaverse

4.2 Blockchain, Non-Fungible Tokens and Energy Marketplaces

Bhujel and Rahulamathavan [6] discuss recent trends, potential applications and related wins and barriers (i.e., safety, visibility and expandability) of the non-fungible tokens (NFTs) as blockchain-based ownership and transaction records across different energy Metaverse marketplaces. The authors argue that due to the

NFTs, Metaverse energy market trading can be made freely available through segregated solutions (DApps, p. 1 above). At a current state, they report that approximately 245 NFTs markets are categorized on multiple assets (products), services and (or) blockchain platforms, representing around 68 million customers (as above). The NFTs footage and dispensation can be processed when data from a blockchain exchange (transaction) is digitally transformed into a suitable format and then placed in an inbound (inflow, internal) or outbound (outflow, external) [i.e., reflecting an open innovation cross-over data access entry, for example [7–13] database based on price tag. Upon signing the deal, the retailer forwards the data to a smart contract (stable repository) that processes it and trades it on the blockchain as transaction. Figure 4.2 next outlines a generic configuration of the NFTs energy Metaverse blockchain marketplace that represents an open innovation-driven agency (capacity), as next.

The above interrelated NFTs blockchain Metaverse market ecosystem "mirrors" the aforementioned internal and external (and vice versa) cross-over linkages between the different agent categories. Stemming from workstation to energy customer(s) through registration, metadata inflow and outflow activity, NFTs creators' and customers' interaction and ending up to ownership pattern and transfer validation, the framework denotes an "open innovation" (e.g., [14–22]) transferable decentralized Metaverse-driven boundary fusion. Unlike physical networks that bear different culture, operations and business objectives, the NFTs energy Metaverse-generated market blockchain is moving towards a renewable-support and decentralized ecosystem with multimodal co-innovator actors (e.g., [2, 6, 23–27]). Having said

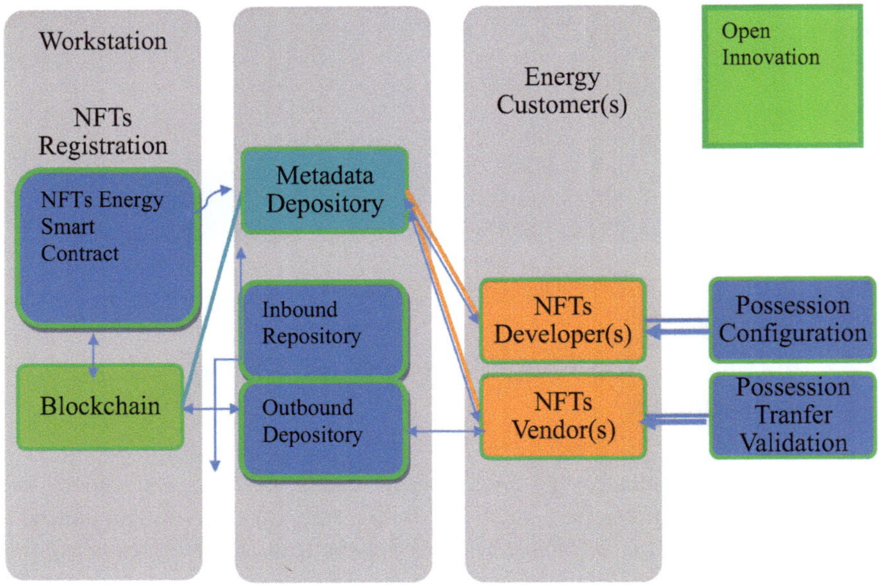

Fig. 4.2 The NFTs energy Metaverse marketplace blockchain open innovation-driven flowchart

that, the above authors [21–27] indicate that NFTs appear to bring forward a rising, yet currently scarce energy Metaverse-led blockchain topic of interest, research, innovation and enterprise (IRIE). The NFTs energy Metaverse-driven corresponding inquiry seems to rest either on safety, scalability and (or) financial-driven issues of blockchain technology NFTs' marketplaces stored in, in general. Moving forward on the aforementioned account and following-up on the authors' urge to reflect upon the energy Metaverse NFTs blockchain market(s) infrastructure, in specific, we reflect upon the prior authors' modelling, properties, benefits and challenges of the energy Metaverse NFTs blockchain market(s) ecosystem, as next.

The Energy Metaverse NFT's Blockchain Marketplace(s) Infrastructure

This ecosystem is envisioned to be stored on blockchain technology in a shared node, distributed, ledger-based network that consists of three tiers. The first one denotes the infrastructure (i.e., network, ecosystem) level (i.e., protocols, links, hardware, etc.). The second one reflects the foundation tier "mirroring" the blockchain itself. The third and last one includes the application (implementation) level [i.e., application programming interfaces (APIs)], human–machine interfaces, smart contract(s), etc.]. Multiple blocks deposit diverse high-secure encrypted data (algorithms). Upon each block data completion, the next block is linked to the prior one creating a chain. The various categories of blockchains appear to store different data types deposited (e.g., buyer, seller, currency, etc.). Figure 4.3 outlines a brief visual mind mapping flowchart of the envisioned energy Metaverse blockchain infrastructure, as follows.

The above energy Metaverse blockchain infrastructure "resembles" an interconnected, decentralized ecosystem of energy network, blockchain and application layers that integrate both energy properties and energy agents' data. The corresponding energy particulars of this Metaverse blockchain ecosystem are briefly presented, as next.

The Energy Smart Contract(S)

This type of energy business deal modus operandi denotes self-performing contract agreement(s) between energy purchaser(s) and energy vendor(s) executed via an interconnected peer-to-peer human–machine network. Such agreements are called smart contract(s) as they empower spread out solutions (Dapps, [6, p.1]). The transferred coded value data storage between the energy agents (buyers, sellers, service providers, etc.) tends to represent a rather safe energy repository process.

The Pace of the Metaverse Blockchain Energy Business Deal

The urge to deliver or buy and sell billions of energy NFTs in a few seconds is expected to constitute the property of speed as vital. The pace of the exchange is considered to be established by the density of the trading and the ecosystem per se. The first energy-based NFTs are stored on the blockchain, the second ones are those priced by the NFT market(s) for each customer (service) provider account and the third of them are rated to register for NFTs for vending. The prior authors emphasize that the NFTs costs tend to vary between blockchain types and marketplaces, resting

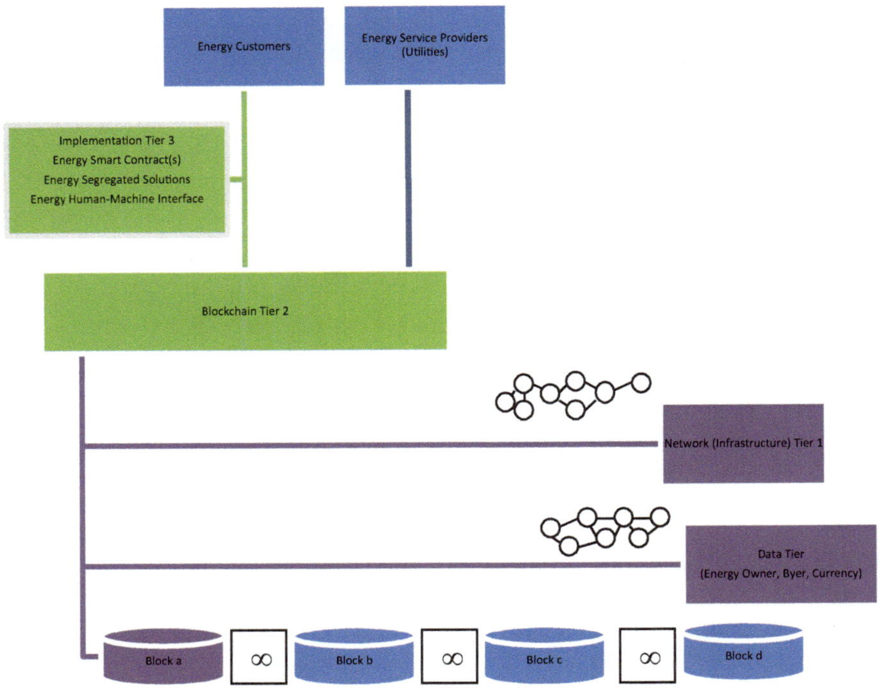

Fig. 4.3 Visual flowchart of the energy Metaverse blockchain network

on the customer and service provider objectives, like the quantity of data employed, the pace of the transaction and (or) the ease of use in each case.

The Level of Metaverse Blockchain Energy Transaction Iteration

Each energy exchange on the Metaverse blockchain is set up in iterative-led blocks, all ending up to link into a chain. The level of iteration can be assessed with different indicators once a block is moved. Pending on increased energy business deal range each day, the volume of the block might not suffice to handle all exchanges delivered by blockchain nodes. Therefore, the blockchain provides with various elements to sideline trading, with one portion (or feature) of the blockchain to be relinquished to hold the others.

The Energy Metaverse Blockchain Tier 1

This tier corresponds to the foundation layer of a blockchain energy modelling. It reflects the baseline architecture of a blockchain infrastructure that aims to heighten the trading aptitude and pace, while accepting more customers (service providers) and data storage.

The Energy Metaverse Blockchain Tier 2

This level denotes an infrastructure that moves along the present blockchain to advance its iterative power, effectiveness and safety, while upholding its decentralized nature. The network is built to manage trading without recording each transaction on the blockchain (off-chain) [28, p. 313: 4], thereby, receiving lowered workload and permitting speedy exchanges. In general terms, Tier 2 runs and reports only a review (synopsis) trading to the baseline blockchain. This route becomes possible by moving some workload of the trading to a nearby structure that manages the volume of the infrastructure dispensation and only gives an account to the foundation blockchain to confirm its outcome, ending up in decreased crossing point on the blockchain's baseline Tier. The preceding authors indicate a number of potential Tier 2 tools, as follows. First, the sideline chains. This type of chains introduces a rather autonomous blockchain infrastructure linked to a further blockchain ecosystem (i.e., core network). The existed consensus protocols (p. 313:2 above) are able to advance the confidentiality and safety of the resultant blockchain and reduce further trustworthiness prerequisites. The baseline notion of the core and secondary blockchains aims to support an easy energy asset(s) transaction. The sideline chains are connected together. They appear not to move any digital energy assets at all. Instead, they are confined on the core blockchain, while on the sideline chains the comparable energy asset is released. Out-off chain transaction is adopted to move the digital energy assets. This process takes place outside of the core infrastructure. Throughout this processing, the trading particulars are passing on to smart contract(s) on the sideline chains for confirmation. Only post-successful confirmation the real money is unconfined on the sideline chains, permitting the digital energy assets to be transferred between both types of blockchains. Second is the shadow chains [29, p. 13]. These chains can manage diverse tasks, are of different types, and are able to sum up the transaction process out-off chain within the smart contract(s), minimizing charges and traffic by expanding the blockchain's trading range and timeline, as well as safeguarding the energy trading data accuracy and validity. Owing to shadow chains, money can be transferred in minutes between various solutions and accounts. In other words, legitimate and accurate transactions are expected to be processed based on less expensive and speedy trading details, in a lower resource-demanding and equal safe way as the primary.

Visibility

NFTs and blockchain share visibility as main feature of their interconnectedness. The latter is in charge of making a recording of all segregated ledger network data on the public domain accessible to everyone. The first are expected to control and make all energy trading properties visible, including wallet assets and possession rights, irrespective of their network. As visibility appears not to be always clear or comprehensible, the principle of visibility related to NFTs is partially showcased (e.g., past NFTs ownership, left over sense of balance, etc.) and usually indicated by the range of accessibility offered by the principal blockchain infrastructure.

Unlimited Accessibility

The authors argue that although various elements might contribute to NFTs visibility, yet the most evident feature denotes that of the community accessibility that blockchain serves across diverse energy cryptocurrency-transactions. The NFT network permits customers (service providers) to cross-check possession and trading records of all energy NFTs delivered up to the latest exchange without requesting any authorization. Following-up on this accessibility option, the customers (service providers) are able to confirm that they have acquired the novel product or service (i.e., energy-related) from the corresponding utility or service sector. The blockchain infrastructure can be categorized into community (public, open), classified (personal, permission-driven) and cross-over (hybrid) classifications. However, the public one is that associated with energy NFTs (e.g., Ethereum). Despite the fact that a number of Ethereum blockchain ecosystems seem to bear a permission-based network for energy enterprise-to-enterprise trading, all relevant data recordings are kept on the baseline community (open) Ethereum infrastructure. In the community (i.e., accessible network), the (energy) customers (or utility providers) might register for any blockchain infrastructure and turn into parts of the shared node network with decreased nodes' privileges. Across this type of blockchain, the shared nodes develop a blockchain for energy trading on an infrastructure, collaborating like a financial entity for a reduced cost. This permits customers (service providers) to overview and confirm the energy trading process. Owing to the extended number of shared node and sized transaction network, permission-free blockchains seem to be safer, but bear a lengthier energy trading process timeline. In general terms, permission-free networks tend to rest on "consensus designs like proof-of-work (PoW) and proof-of-stake (PoS)" [29], pp. 1–2] and maintain data (recordings) secure from breaking through. Permission-free blockchains have spread extended usage for energy NFTs across diverse tools. This type of technology seems to permit (energy) customers (utility providers) to trade corresponding energy NFTs without compromising their identity or go through client monitoring that might need administration confirmation. In this way, the (energy) customer (utility provider) and their energy NFT asset can be kept on the private sphere and keep on doing so. The reason for this is that when an energy trading is performed, the transaction details (e.g., the customer's wallet, the transaction's details, etc.) are registered on the blockchain. In case the wallet's identity (ID) is not connected to that of the customer's, the (energy) trading procedure residues unnamed. A few number of existed virtual NFT markets seem to offer the opportunity for customers to buy (sell) diverse product or service NFTs assets for cryptocurrency. We might see relevant energy NFTs marketplaces whereby energy customers and (or) utility providers might store (list), display and sell corresponding energy NFTs on the public domain within minutes in the near future.

Permission-Driven Accessibility

Against community (i.e., open) blockchain infrastructure, this category demands a permission-based blockchain where users (customers) are monitored, data is handled, and energy trading is confirmed. Although a segregated infrastructure with a dispersed structure, this ecosystem is monitored by one partner institution deciding on the node(s) context and properties. The emphasis rests on iteration aspect, removing the space or option for fully-fledged visibility, as the authority organization controls each use case validation. The authors conclude that each type of blockchain energy networks are characterized by their wins and (or) barriers, the community (open) ones segregated at the cost of pace, for instance, and the permission-driven (personal, private) ones focal, but robust.

Energy Trading Record

The process of energy trading verification appears to be clear-cut. Resting on which blockchain network the (energy) customer (or service provider) accesses to acquire, vend and (or) generate energy-based NFTs, he (she) can get data from a number of resources. All related transaction recordings (i.e., energy NFT acquisition, vending, request removal, and remuneration data) are expected to create an exclusive blockchain identity energy trading. Everything tends to depend upon the cryptocurrency-based energy NFT acquisition. For instance, Ethereum, one of the most widely used blockchains is tied in with tracking capacity and multiple seeking out preferences (i.e., wallet location, trading jumble, pay slip ID, etc.). At the end of the day, it seems to be all about selection and what the (energy) customer (or utility provider) is seeking to achieve pending on his (her) objectives. That is, a network that can offer quicker exchanges, a fund energy NFT trading, a superior size (range) of energy NFT dealings, and (or) a more powerful safety-led energy trading process.

Safety

Despite the NFTs marketplaces developments following-up recent interest, the authors note that not all NFTs appear to be entirely safe. To begin with, it seems rather vital to emphasize that ill-intentioned customers might be ready to take advantage of any valued energy assets immediately. The attractiveness of NFTs has turned out to be very widely known, so this catches the attention of potential intruders to explore NFT weaknesses, which can considerably challenge the safety of energy NFTs. For example, it is common knowledge that a NFTs market customer (service provider) should never disclose or share with anyone his (her) private key which includes the (energy) assets to exchange, the proof of possession (purchase) and (or) verification. Along these lines, the 2-factor authentication followed in different recommended energy solutions aims to protect the customer account to be compromised. Accessing the collection of NFTs digital energy assets may well pose the first vulnerability step for intruders to take. Since cryptocurrency is not a physical but a digital asset, it appears to be challenging to pin down (during mostly anonymous energy trading), when transformed from money to digital. As such, it might be trickier to safeguard from cash-laundering. A centralized-led energy supervisory body overseeing this activity could further support the protection of energy NFTs marketplaces from criminals.

NFTs Repository

Various file systems are used to store NFTs on the blockchain and safeguard their security. Among others, hashing algorithm data indicators (HADI) are diametrically linked to the NFTs domain, crafting the NFTs content as one of the more secure modes for NFTs storage, while contrasting Hypertext Transfer Protocol (HTTP) addresses being vulnerable to amendment and intrusion. In addition, a HADI can only be one-content related, so (energy) customers (or service providers) can be certain that no modification can be made without breakage-link, thereby, setting it as rather optimal for energy NFTs stockpile. Further, smart contracts can be employed to revise the address of an order by permitting the change of it, while storing a trace (proof) of the primary account in the blockchain energy trading domain to guarantee everyone is updated on any modifications performed. In sum, HADI seem to bring a safer selection when energy NFTs repository comes into mind, however, they appear to be rather susceptible to intrusion attempts.

NFTs Wallets

The NFTs are expected to be stored in a high-performing safe space to protect their value. This storage can be applied through cryptocurrency wallets [30]. Such wallets reflect either device or software that permits (energy) customers (service providers) to reposit and exchange their virtual energy assets. Such operation is configured to operate on the blockchain network that NFTs are set up and might be supported by (energy) customer (utility provider) services. In addition, they can embrace both public [i.e., Internet Protocols (IPs) on blockchain] and private (i.e., proof of possession of remuneration) keys for asset transmission (pp. 1–4 above). As they are built on the blockchain and rather not in the energy customer's billfold, they can offer the energy consumer with ease of use when dealing with (energy) cryptocoins and NFTs blockchain transfer. Their configuration begins when seeded phrases (i.e., array of terms) are produced, at first.

These words are then employed as the private keys' basis, but they should not be made known to anyone in order to avoid unauthorized users (e.g., customers, service providers) to access the linked digital assets. Irrespective of the nature of their type (i.e., either hot or cold ones; [6, p. 12], the (energy) NFTs wallets can be set up in any machine. The hot ones (also known as software wallets, as above) can reposit the proofs of (energy) acquisition funds on the web. This long-term online storage seems to facilitate the marketplace energy trading, however, it seems to heighten the risk of unauthorized access (intrusion attempts). The authors note that against high-performing safety services, web wallets continue to pose a rather challenging threat for attackers and the best way to address their vulnerability might be to ensure assurance. The cold wallets (also popular as computer wallets) tend to include a memory stick that stores cryptocurrency keys and NFTs. With a computer wallet, the (energy) customer (utility provider) can access his (her) energy digital assets on the device (not online), making the energy transaction(s) less vulnerable to intrusion attempts. Even if the hardware wallets were online, the authorization of the energy trading would take place on the machine and then it would be transmitted to the blockchain for NFTs. In this sense, the proofs of acquisition of energy funds will not be removed from the device and there would be lower chances for attackers to fake any signature. Therefore, the cold repository wallets are considered to be the best option for safe and impassable (energy) NFTs storage.

Blockchain Networks for NFTs Markets

The above authors emphasize that selecting a blockchain infrastructure for energy NFTs marketplaces might be very challenging, given that energy NFTs denote a non-fungible property or pillar, that is, previous digital energy assets do not become visible on layer blockchains. The different types of blockchain ecosystems bear various properties, wins and challenges. For example, the Ethereum seems to be the prevalent one in terms of volume and popularity, offering segregated financial and legal service, and supporting the creation of smart contracts. However, it comes with increased energy trading costs and power running, due to its PoW operation (p. 7 above), requiring machines to race each other to resolve difficult tasks in order to insert blocks and trading interventions to the blockchain. When switched to the PoS operation (networking) (as above), the Ethereum allows the network to experience better scalability and energy trading speed, while limiting the costs run. The authors emphasize the fact that the switch to the PoS energy networking appears to have decreased the infrastructure's power expenditure by more than 99%, thereby, establishing it as a more power-effective blockchain that seems to promote environmental sustainability without compromising the Ehtereum's velocity-and-scalability-driven properties. In another instance, the Solana blockchain seems to be characterized by advanced performance in terms of speedy transaction, associated cost and power-efficiency reflecting an eco-innovation commercial mode. This type of blockchain network is considered to combine the proof-of-history (PoH) operation (networking) with that of the PoS (p. 13 above). This hybrid combination tends to offer faster transaction, accurate orders and appropriate tracking of energy trading supported by safety paradigm. Despite its stable and secure network, the Solana blockchain seems to lack extended adoption, therefore, it is expected to host limited NFTs energy marketplaces and connected energy-driven customers (or utility providers). Although it tends to be novel for the time being, it is expected to rather increase its promising power among the accelerating blockchain networks for NFTs marketplaces and, optimally, those built-up on energy markets, in particular. An additional example of blockchain network reported for NFTs trading is the Polygon (p. 4 above). This kind of multiple-level chain infrastructure lets new products or services to link to the compatible Ethereum blockchain, taking advantage of its scalability and safety properties. Its associated costs pertain not with the NFTs creation, but with their sale. Nevertheless, the Polygon seems to be rather restricted in range, at least for now. However, the authors suggest that both blockchain networks, that is, the Ethereum and the Polygon should be treated as merged ones, with synergistic impact, and not as competitor ones. This can be made possible since the Polygon can link to the Ethereum's interoperability, elevated trading speed, limited costs properties, and introduce new scalable tools. The Binance Chain which runs along the

Binance Smart Chain network (p. 9 above) is indicated to offer the Ethereum additional support over the smart contract energy trading. This makes the Binance Chain and the Binance Smart Chain networks as potentially effective Ethereum substitute solutions. Their speed and trading pace are advanced, with limited costs. However, their operation tends to be central, thereby, rather more susceptible to unauthorized access and intrusion attempts. The Binance Chain and the Binance Smart Chain network(s) connected PoS authority operation (pp. 13, 16 above) seems to garner certain competitive advantage(s) for the corresponding (energy) NFTs marketplaces (e.g., "restricted deny access to" time, reduced costs, etc.). In that respect, the prior authors argue that this category of blockchain networks for NFTs are expected to embrace all Metaverse-driven (energy) different ecosystems combined. Last but not least, the Cardano network is reported as another kind of blockchain infrastructure for NFTs marketplaces. Upon its' PoS operation launched, it ended up one of the most widely used blockchains on the NFTs marketplaces. Originally developed by blockchain industry expert teams, it bears classified scalability and power-efficiency, making it a rather eco-sustainable blockchain network for energy NFTs markets. Likewise the Binance Chain and the Binance Smart Chain network(s), Cardano employs Ethereum modelling to facilitate corresponding smart contracts and application programming interface(s) (APIs) (pp. 16–17 above). The API for running on the blockchain is reported to be offered as a cost-free community support, permitting (energy) customers (or service providers) to interact promptly with the shared node(s) or consumer(s) networks.

Energy NFTs Marketplaces

NFTs markets can store, share and trade collectible digital assets of many product (service) and form transactions (e.g., smart contracts, artwork, energy, screenshots, documents, audio and video files, etc.). This process is made possible through a blockchain ledger encrypted network that safely deposits, visualizes and handles currency and NFTs. Across the blockchain platform, the digital verification of energy good (service) possession, in particular, is signified by token-based representation for all NFTs digital energy assets. The previous authors argue that the NFTs energy infrastructure consists of three sub-parts, as next. First, the content (energy product) developers. Second, the vendors (energy suppliers) and third, the consumers (energy purchasers). The first ones build digital energy asset material from primary data and locate it to the host service to be available for public use. In cases where the creators do not transfer their content into an energy NFT to be able to trade it on the blockchain, the energy traders can create a unique token on the latter (i.e., "mint" the energy NFT for transaction) (p. 6 above) [31], pp. 2–3]. After registering the digital energy assets on the NFTs marketplaces (decentralized), energy customers (or service providers) can bargain any selected energy good (service) at a given cost. The energy consumers (or suppliers) may deliver energy trading through the NFTs web application platform to a smart contract that executes the NFTs' and (or) blockchain's set of rules for formatting and processing data (i.e., protocol), permitting them to various exchange activities. The latter can be performed either on-the-chain, off-the-chain and (or) in a hybrid chain mode [6, p. 20]. If deposited on-the-chain (p. 19

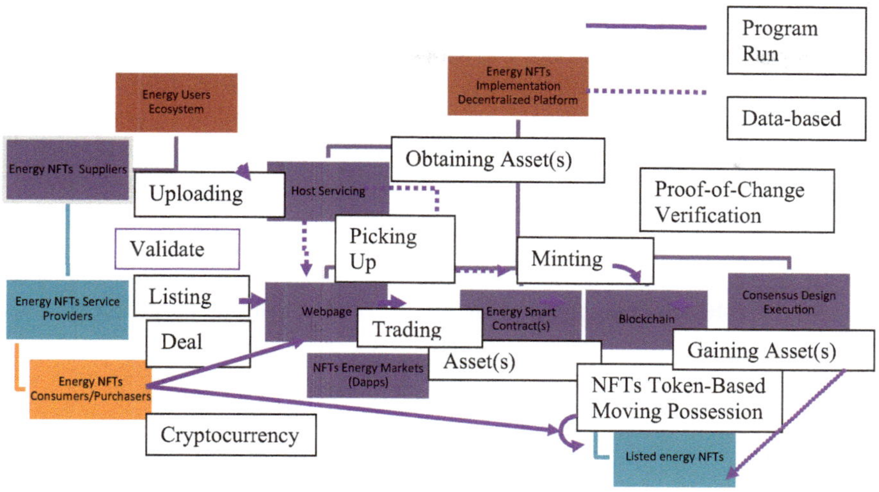

Fig. 4.4 Flowchart of the open innovation-driven blockchain energy NFTs trading network

above) the energy trading tasks are streamlined on the blockchain and are signature-based authorized, with increased operating costs. In case of off-the chain (as above) repository, all energy trading activities are logged into cloud in an encrypted-based mode and then deposited on an off-the-chain platform and not on the blockchain, utilizing decreased operating costs. In the event of the hybrid-chain storage, a combination of on-and-off-chain deposit is expected to ensure decentralized, transparent and secure protocol(s). Following-up and expanding upon the authors' aforementioned account of blockchain networks for NFTs marketplace(s) transferred as those for the energy NFTs market(s) ones, in particular, Fig. 4.4 that follows sketches a flowchart of the open innovation-led blockchain network(s) for energy exchange NFTs marketplace(s), as next.

As showcased above, the energy service providers or suppliers might develop energy (digital) assets and transfer them to the host servicing. This service runs as an outflow (i.e., outbound, external) database outside the blockchain (i.e., inflow, inbound, internal database) that sets the energy-based assets accessible to clients. The fusion of the aforementioned service as an interrelated inflow and outflow boundary energy access platform can be transferred as representing a cross-over open innovation entry network (e.g., [14, 32–43]). Within this infrastructure, if the energy service providers lack the technical capacity to renovate their energy product (or service) into an NFT and vendor it on the blockchain, the marketplace(s) can offset this discrepancy by allowing vendors to energy "minting" for trading (see above), but energy suppliers can also trade NFTs if they bear the prospective capacity. Following the listing of the energy assets on the decentralized markets, the consumers (clients) can proceed with energy bids (offers) and buy the energy assets at a designated cost. The energy digital assets' consumers are able to forward dealings through the website application to smart contracts that employ the NFT marketplace formula and blockchain, along with a token-based contract that ensures various related energy trading duties. The latter can be garnered either on-the-off-the and (or) the-fusion chain. If stored on-the-chain, all tasks that are connected to the NFT marketplace(s) go "live" (p. 19 above) on the blockchain, with the token-related energy asset(s) certified in a signature-based mode and possess increased cost due to the on-the-chain modelling. If operating off-the-chain, all activities are documented on the cloud (i.e., centralized platform) in an encrypted form. The encrypted type data can then be deposited on the-off-the chain application and not on the blockchain. This operation can ensure a less costly function for the user(s). When all tasks aligned with the energy NFT marketplace(s) are located on the fusion-chain, cryptographic-based monitoring offers a decentralized and safety net for engaged energy agent(s).

The Energy NFTs Marketplaces Challenging Issues

Safety

The aforementioned authors indicate that there seems to be a reported record of security challenges associated with the energy NFTs trading that need to be taken into account. For example, in recent years, the number of potential fishy NFTs registrations has risen by almost 300% (as above). In addition, extra attention has to be allocated to safeguard possible suspicious access to energy customer (or service provider) NFTs crypto-wallets accounts. Another relatively recent malicious attack incident ended up in stealing $2.9 million worth NFTs tokens (as above) from an existed NFTs marketplace. The prior authors describe the corresponding intrusion attempt, as follows. First, the attacker put his (her) signature on a fractional contract, leaving behind a big section empty. Following the NFTs signature, the intruder finished the contract by a call option which conveyed the possession of the NFTs without any expenses. Although such attacks are popular, there tend to be multiple tools to moderate similar risks. Having said that, the associated NFTs marketplace advised its customers to drift NFTs from the operating at that time Ethereum blockchain to a newer smart contract (p. 20 above). While this upgrading service was taken place, the attackers embarked on a refined hacking movement, permitting them to take NFTs from the repository space. Once the inquiry was over, it was revealed that the intrusion attempt was launched on the outside of the specific NFTs marketplace. The preceding authors emphasize that any energy NFTs marketplace should be expected not to interact with energy customers (service providers) by email, but only when to enrol them. The use of cold wallets across NFTs storage is further recommended as additional measures to mitigate security risks. In other sophisticated malicious attempts, a group might extract a venture's tokens, vanish with the money, and leave the shareholders with valueless digital energy asset(s). This can be made possible either when developers insert harmful code to NFTs tokens to impact their worth, or in cases where the content creators trade great volumes of NFTs tokens or put them on sale in parts, constituting them valueless. Resting on the free nature of the NFTs marketplace, the developers are not considered to trade their NFTs tokens unlawfully. This seems to make it more challenging to spot them. The reporting authors advise for energy account(s) cross-checks, taking extra care for decreased energy transaction range, increased liquid-driven activity, analysis of the energy acquisitions made, and (or) inspection of the wallet-based values, among others, to leverage protection from prospective threats. In the case of NFTs marketplace off-chain storage, the fully-fledged data and energy trading recordings are deposited off-chain, making it challenging to validate the associated energy transaction or possession account. A harmful NFTs marketplace might exploit this condition by falsifying forged selling recordings. Having said that, the authors end up with the question whether the organizational entity behind the energy NFTs marketplaces ensures authorized access or not. In addition, they recommend the following reflective queries asked pre-buying any energy NFTs venture, as next. First, re-consider the core 2 distinctions that are present between the blockchain networks for NFTs marketplaces. The public decentralized ones tend to be characterized by advanced security. Their fully-fledged transparency appears to make them less susceptible to be compromised. The private ones endorse energy customers (service providers) sheltered from unsolicited entrance, the energy services offered are secure and energy trading costs are limited. No unlawful attempts are allowed, with rules (regulations) plainly run through. The previous authors argue that in both blockchain networks for (energy) NFTs marketplaces, each one bears different advantages and disadvantages. Thus, it tends to be rather a matter of energy customer (service provider) selection at the end. They note the case of content creator royalties in smart contract NFTs marketplaces as well. They claim that although the smart contracts seem to permit energy customer-to-customer (or service provider- to- service provider) trading, they tend not to be enabled across diverse blockchain networks. Thereby, the prospective creator royalties run on one NFTs marketplace appear not to be detectable to others. Building upon this articulated discrepancy, malevolent vendors (suppliers) could escape royalties by exchanging energy NFTs on spaces with no set up royalties, with the content creators being unable to avert purchasers from registering the energy NFTs on new marketplaces, ending up in the vending of each energy NFT and the loss of these energy transaction royalties. Despite the fact that NFTs royalty costs seem to be embedded into the NFTs smart contacts, their inclusion nature seems to be optional, as there tends to be a mode to discern between energy trading transfer and (or) auction, with mere off-chain smart contracts energy NFTs marketplaces paying energy content creators' royalties. In this regard, unsolicited vendors might transfer the energy NFTs to the energy purchaser to prevent this fee, run the imbursement off-infrastructure or being charged, setting the software interface (SI)-transfer energy smart contract more costly. The prior authors suggest the so called "universal token standard" (p. 22 above) as available tool to mitigate any energy royalties challenges. According to this solution, the energy content creators can maintain their royalties and have them acknowledged across the full range of the blockchain network(s) for energy NFTs marketplaces. This is expected to confirm that their energy creation is credited. Novel energy smart contracts can support advanced energy trading, such as the subscription-led one, which can be employed to assign forthcoming vending of energy NFTs to energy content creators. Another recommended option might be to advise the latter to integrate guidelines on their energy NFTs smart contracts to suspend digital energy assets to penalize royalties-driven frauds. In terms of the blockchain network for energy NFTs marketplaces repository, the authors note the scalability-driven challenges endorsed by the provision of the on-and-off chain storage, empowered by the expenditure of sustaining great volumes of on-chain database(s) upon the infrastructure, operating over and over again, making it more costly, eventually. In cases where the cost of each energy trading heightens, so does that of the energy deposit, with a number of energy NFTs marketplaces using the off-chain repository operation. In this case, the digital energy assets data linked to the energy NFTs on the blockchain are stored in a way that showcases that they were designed genuinely by their creators. This type of repository appears to exercise a scalability-based and cost-effective storage, but it might entail transparency shortage, as well. The authors propose the implementation of energy smart contracts to deposit energy (transaction) records in an infrastructure of energy repository led by energy customer-to-customer (service provider-to-service provider) distribution (e.g., various copies of transaction records). In the case of one repository workstation being inaccessible, the records can be still accessible somewhere else. Other storage options might be to deposit data in a repository platform purchased by the energy customer (service provider), the cryptocurrency cold wallets as the safest mode of maintaining energy NFTs and metadata "off-the grid", password-protected and thereby, indissoluble, and (or) the off-chain deposit for energy NFTs, minimizing hacking risks. The above authors further stress that the blockchain networks for energy NFTs marketplaces advanced software development and engineering power, demands and data storage costs, appear to constitute even the simplest energy trading costly. For example, the Ethereum energy NFTs customers (or service providers) need to pay advanced costs to buy NFTs on the infrastructure, with most blockchain networks bearing elevated NFTs transaction costs, in certain cases higher than that of the NFTs value. In addition, while the associated costs are increasing, the expansion of novel NFTs markets is decreasing. Given that elevated fees are haunting the NFTs marketplaces, the authors note that a number of blockchain networks have begun to operate on Tier 2 protocol applications or abandoning the Ethereum blockchain for another candidate(s) infrastructure. Some other alternatives might include the usage of Tier 2 system solutions on a regular basis or to select the lower transaction cost per timeline, but not that of the slowest speed. Expanding upon the aforementioned authors' recommendations and rationale, Fig. 4.5 outlines a summary (overall) visual conceptual model of the potential challenges that Metaverse blockchain network(s) for energy NFTs marketplaces might entail, as follows.

Exploring the notion of an envisioned smart intelligence, sustainable (eco-friendly) and safety-run Energy Metaverse infrastructure, the following authors [23, 25, 27, 44] argue that the new and rising Metaverse-driven emerging technological advancements seem to lead immersive technology towards innovative energy production. They report that the accelerating demands for sustainable-based energy and the appearance of Metaverse have offered the chance to reflect upon how to develop safe and sustainable-led energy trading ecosystems by implementing diverse and challenging intelligent technologies [e.g., smart grid(s) (SGs), virtual power plant(s)(VPPs), digital twin(DT) and blockchain (B), among others]. The above authors suggest that assimilating the VPs technology, for example, with that of the

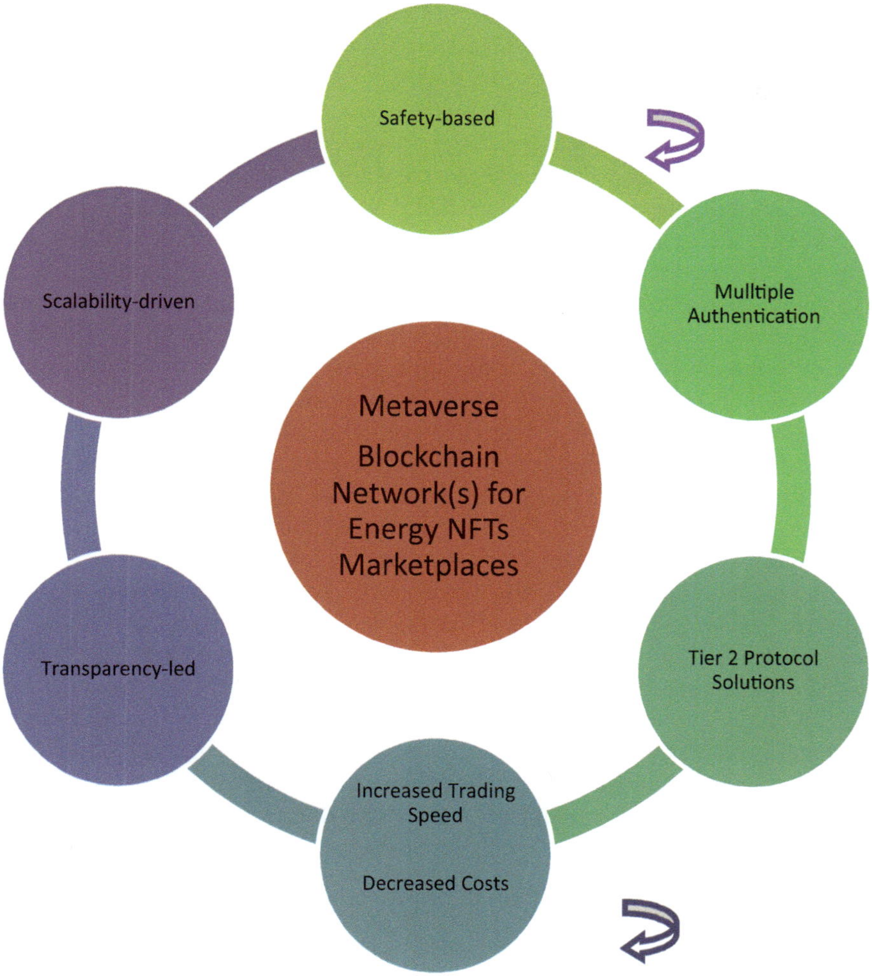

Fig. 4.5 Metaverse blockchain network(s) for energy NFTs markets challenges

SGs might improve the exploitation of "distributed energy resources" (e.g., [27, p. 2]. That is, energy systems generated by renewable resources are able to increase the Metaverse's steadiness and trustworthiness of its energy supply chain management. In addition, they note that the application of the DT technology might support the effective running of the VPPs by replicating their functioning and properties to locate possible challenges and screen diverse use cases for development. Further, the B technology is recommended as another emerging strength that could facilitate disperse and reliable solutions, able to generate Metaverse-based distributed energy transaction network. Resting on the above rationale and argumentation, the authors suggest a theoretical and practical infrastructure ("Next-Power") [45, p. 2015] that might be used to generate a secure and sustainable, Metaverse-led energy trade ecosystem. This ecosystem is considered to integrate the following parts, as next. First, the energy trading blockchain-based compensation between customers keen on consuming and purchasing energy. This activity can be managed through the smart contracts on the Ethereum B network to set the energy trading fee by employing a collaborative gaming approach (i.e., overturn sale). Second, another tokens' driven B infrastructure that can enable a safe, shared, authentication-based process between clients involved with energy exchange. Third, a less demanding on the B plan that can monitor the energy customers profile on Metaverse. Reflecting on the authors insights, their recommended ecosystem is applied, delivered and assessed on the Ethereum B formal testing infrastructure. Overall, they indicate that their framework tends to show a potential practical, adaptable, effective, secure and financially viable approach that could support the "alternative sustainable future" of Metaverse-led energy trading. In sketching their inclusive contribution, we may argue that their proposed framework is an innovative-led one. This innovation seems to leverage the smart contracts on Ethereum B technologies to assemble a safe, visible, adaptable, resilient to cyber attacks, of decreased expenses, and trustworthy peer-to-peer energy Metaverse ecosystem (p. 2022 above). In addition, their suggested infrastructure exercises open-ended competitiveness and legitimate Metaverse energy trade run on "on-the-chain" [6], p. 19) smart contracts (i.e., rolled-over sales) and "off-the-chain collaborative game theory" (p. 24 above). Reviewing their insights, we envisage that they also designed a digital entry token-driven B network that enables the off-the-chain shared verification-based process between the VPPs clients that buy and sell energy. In the latter, they have generated a less intensive on-the chain system that tends to screen the customers use cases profile in a fully-fledged de-concentrated mode. Lastly, their ecosystem performance appraisal experimental assessments seem to showcase the Metaverse-built adaptable, safe, lowered-cost and eco-effective strength (capacity) of their energy trading modelling approach.

The Proposed Novel-Energy (Power) Framework

Envisioning the above author recommended inclusive energy trade modelling framework, we claim that it tends to integrate the following core parts, as next. First, the suggested on-the-chain (see above) programmed, visible imbursement process used for energy trading. This workflow aims to ensure that expenses are dealt with in a time-and-transparent, cost-efficient, one-to-one trading, without any intermediaries engaged. Second, a legitimate and open-ended peer-to-peer energy disperse marketplace that leverages nourishing antagonism and decreases expenditure. Third, the off-the-chain (as above) shared verification process resting on token-access, lined-up to support safe contact between engaged agents. This paradigm aims to safeguard that all parties involved are authenticated-based agents in the energy trading, with no pending on-the-chain-trade instances. Forth, the on-the-chain (see above) cooperative trading profile that targets at fostering a disciplined and of good habits transaction, without any energy marketplace discrepancies occurred. This engaged behaviour can establish a reliable use case and profile record in the long-run, supporting the stable and robust status of the energy marketplace(s). An outline of the prior authors projected novel Metaverse Ethereum, smart sensing, and renewable-driven smart contract energy trading ecosystem is outlined in the following Fig. 4.6. We operationalize their recommended energy trade modelling as an external (outbound, outflow) exercise of renewable resources, smart sensors (SS) (metres), virtual energy factories (VEFs), constructions (companies) and houses to be implemented on the internal (inflow, inbound) Ethereum B, facilitated by smart contract(s)(SCs). This outflow-to-inflow energy trading resources (i.e., agents, assets, particulars) entry (embracing) represents an open innovation pattern (e.g., [46–53]). More specific, the architecture of the framework depicts the 3 core units of their ecosystem, as next. That is, the Ethereum B infrastructure, the energy customers that can buy and sell energy through disperse renewable energy resources (RERs) (e.g., photovoltaic panels, wind generators, etc.) and the safe network that ensures the trade of excess energy. Each energy agent smart house appears to operate through smart sensing devices that support the Ethereum ecosystem by lending actual data record on energy generation and expenditure. This piece of data seems to be employed to streamline and serve the legitimate and visible energy exchange, safeguarding that the energy players are remunerated for their offerings to the energy marketplace(s). The Ethereum B infrastructure performs as a safe, disperse (non-centralized), visible, concrete and impenetrable network for handling peer-to-peer (as above) energy exchange. The full range of processed energy trading(s), expenditure and profile status operation(s) are executed through a SC framing. This aims to safeguard a safe, reliable and visible energy marketplace for all energy agents. The SCs sub-network comprises four types of SCs, as next. First, the managing (running) contract. Second, the sale one. Third, the interconnection contract and forth, the trading profile entity. The contracts operate in an automated way on each intersection of the Ethereum B infrastructure, reducing the demand for a central platform to handle energy trade and expenditure. This distributed approach seems to enable a high-performed level of safety, fairness, and visibility in the "peer-to-peer" (p. 22 above) eco-friendly energy marketplaces.

In terms of the overall SC energy ecosystem recommended above, the authors have forecasted it to safeguard contracts code reprocess and abridge their deployment. Their projected architecture rests within the Ethereum B and is employed to handle peer-to-peer energy trading between clients who buy and sell energy on Metaverse. These customers tend to own the decentralized renewable energy resources that generate energy. They are further supplied with smart sensing devices that control and convey energy production data recording to the Ethereum B through the novel infrastructure. According to the author's insights, the SCs aim to set an automated-based energy trading procedure and reduce the demand for a central network to handle energy trade and expenditure. On the other side, customers are expected to use smart sensing devices that can support automated everyday energy activities (e.g., invoices, screening, scheduling, etc.) through the novel SCs energy infrastructure. This architecture seems to facilitate clients to join the peer-to-peer energy marketplace(s) and be able to get energy at a decreased fee from other customers who purchase and possess extra energy to vend. Figure 4.7 illustrates an outlined overall account of the aforementioned author's interrelated SCs energy network integrated into their novel infrastructure, as follows.

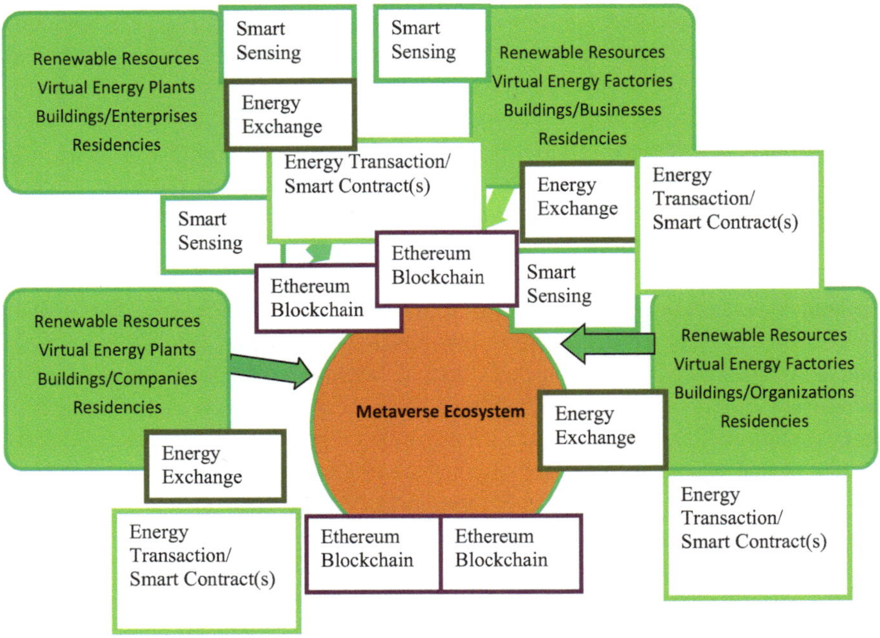

Fig. 4.6 A novel open innovation-led Metaverse ethereum-smart and renewable energy trading infrastructure

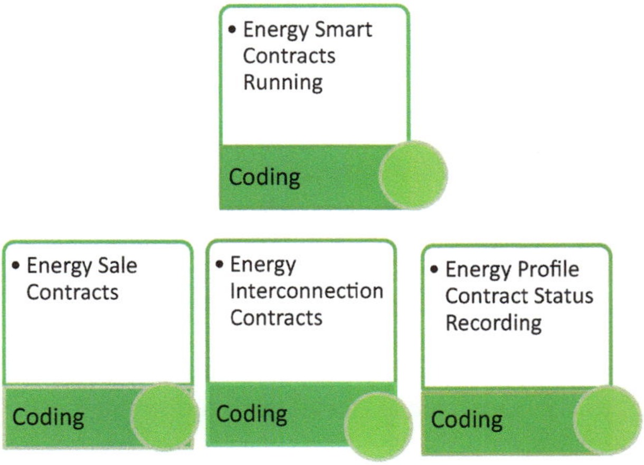

Fig. 4.7 The author's novel framework-led interconnected smart contracts energy ecosystem

The aforementioned inclusive, novel-based, cross-over, SCs energy network might be further categorized to represent the following ranked (i.e., intervening associations between the SCs particulars) Fig. 4.8 flowchart that "mirrors" the novel, energy smart contract sub-entities. Similar to the prior reported open innovation, Metaverse

Ethereum, smart, renewable energy trading infrastructure, B energy NFTs trading network, and energy NFTs Metaverse marketplace B flowcharts, Fig. 4.8 sketches the SC Energy Buying (and Selling) Infrastructure that aligns with the open innovation-powered paradigm. This modelling pattern (e.g., [54]) is supported by the fusion of the outbound (i.e., external, outflow) application (or access) of the energy suppliers vending along with that of the clients buying behaviours into that of the inbound (internal, inflow) interconnected SC energy trading fulfilment and energy sale SC bids. Notwithstanding the bidirectional interrelationships existed between the energy contracts transaction particulars (e.g., token-based access, compensation, etc.) and those of their status pipeline.

Reflecting on the above author's rationale, we claim that they argue that any participant institution that owns the SC running energy scheme (e.g., energy service or product provider) could develop the operation and embed numerous approved members (i.e., decentralized energy customers and suppliers that buy and sell energy) into a mapping-generated mode. This mode includes all energy tender(s), the running SC energy system and the sales SC metadata-based energy record(s). Each offer tends to be connected to no more than one decentralized SC that generated the energy sale

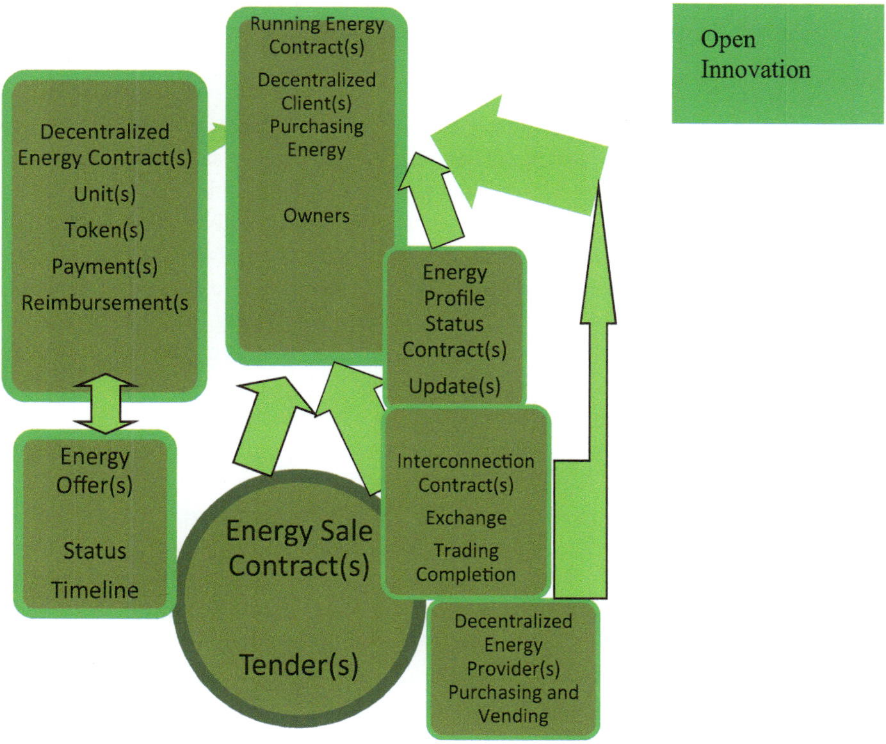

Fig. 4.8 The author's novel open innovation smart contracts energy buying and selling intertwined status ecosystem

and garners data about the lowest-associated bid-energy account(s). The interconnected SCs manage the linkage(s) between involved stakeholders, keep record of any payments endorsed and initiate the token-based entities that are employed for shared off-the-chain verification [6, p. 8]. Each participating energy-related agent is connected to the corresponding energy-linked profile status scores. The latter are based on the profile status process. This function endorses the operation of the running energy SCs and handles the profile of the decentralized energy suppliers and the accredited standing of the similar energy consumers.

The Running Energy Smart Contracts

The prior authors argue that the Metaverse-led energy trading ecosystem is created to handle the peer-to-peer energy buying and selling among diverse energy users (e.g., both energy consumers and energy suppliers). Any business can first develop a SCs-driven architecture for their infrastructure and set it up on the Ethereum B to safeguard visibility and safety. Through the running SCs, the company may integrate the energy users' related data to the peer-to-peer auction network [e.g., Ethereum Internet Protocol (IP) addresses, preliminary energy profiles information, authentication accounts, etc.]. This process facilitates the straightforward inclusion or exclusion of energy users and supports the decentralized and visible running of the peer-to-peer energy auction ecosystem.

The Sale Energy Smart Contracts

This kind of contracts is further suggested by the preceding researchers as helpful one in buying and selling energy goods and services at a reduced fee, safely. This approach seems to associate multiple energy product suppliers with possible energy purchasers (i.e., clients, consumers), facilitating a rather more cost-effective, eco-friendly and smart intelligence energy auction due to flexible, legitimate and open-driven, energy price bargaining. During this operation, the customers generate an energy demand, with the accredited energy product or service suppliers offering their bid for the fee at which they are more likely to auction their excessive energy. Once the energy sale ends, the energy producers with the minimum energy offer win the auction and re-order the role of customers and suppliers from a frontward mart (i.e., A-D below). The following Fig. 4.9 outlines the above auction energy-linked, SC sub-system, as next.

Along the same lines, the interface between the decentralized energy clients (consumers), energy product or service suppliers and the SCs ecosystem in an overturn sale (auction) mode are sketched out in Fig. 4.10, as next. Following the energy owners who embed the paradigm of authorized players (i.e., decentralized energy consumers and those of energy product or services suppliers), the first generate the sale (auction) by paying in a particular sum of money to guarantee legitimate exchange. The decentralized energy customers are tasked with stipulating the quantity of energy demand, the offer timeline, the preliminary fee, and a least standing worth for the decentralized energy product or service producers. The latter begin to forward their bids (i.e., corresponding to the amount at which they are expected to

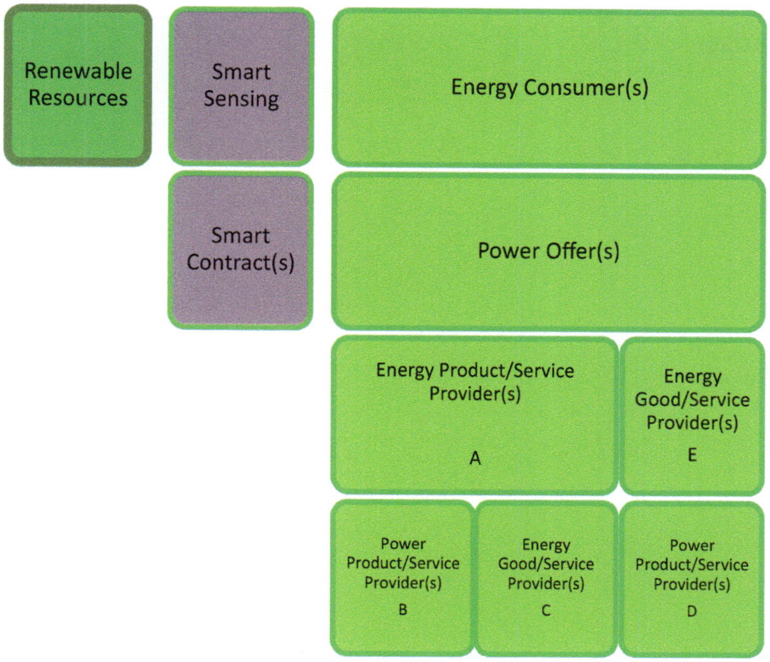

Fig. 4.9 The energy sale (auction) smart contract overarching sub-network

trade their excessive energy) to the sale (auction) SCs energy interaction infrastructure. In case a decentralized energy product or service bid is received, an equal sum of funds is kept as a credit, along with the minimum standing worth and the bidders account data for additional assemble. If not, the bid is expected not to be undertaken. When the bidding-timeline is completed, the decentralized energy consumers may finish the sale (i.e., auction) by generating the contracts operation. The money for the energy vended is estimated and the inbound (i.e., internal) operation of the interconnection contracts is set upon to calculate the requested tokens for the off-the-chain shared authentication running. Once the decentralized energy product or service suppliers have offered the requested energy to the decentralized energy consumers, the latter end the linkage and charge the decentralized energy suppliers according to its use. The SCs pass on the funds and secured payment to the decentralized energy product or service suppliers, and those left behind counterbalance to the decentralized energy consumers, closing the interconnection SCs network (i.e., VEFs). In the following Fig. 4.10, a diagram of the preceding author's energy sale (i.e., auction) and mind mapping life-cycle showcases a comprehensive account of the corresponding network.

The Interconnection Smart Contract(s)

This type of SCs initiates the "access tokens" (p. 24 above) that support the safe off-the-chain interaction between the consumers and the suppliers that provide with the

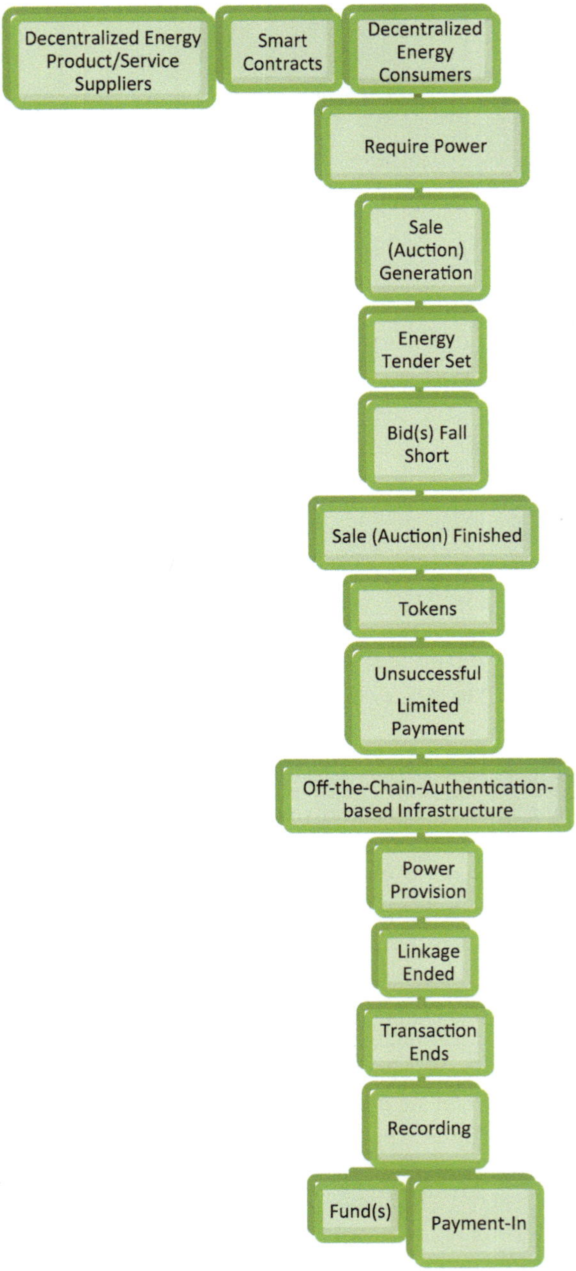

Fig. 4.10 The energy sale (auction) life-cycle diagram

lowest fee for energy. The tokens are employed to maintain a safe and reliable linkage between the two agents, permitting the removal of trusted data such as credit details and energy spending information. The interconnection contracts first authenticate the consumers and suppliers based on their wallet-associated IPs and keys-in. When both connection partners are verified, the contracts initiate tokens that are exclusive, encrypted and safely transferred to the consumers. The latter are able to utilize these tokens to interact with the chosen suppliers and generate an energy trading. The tokens tend to ensure that only verified partners are expected to be involved with the off-the-chain linkage, as protection means against suspicious attacks on the established connection. By and large, the interconnection SCs appear to be a core element of the authors recommended novel SCs energy infrastructure, ensuring safe and effective off-the-chain interaction between energy consumers and energy suppliers in the one-to-one engaged energy marketplace(s).

The Profile Smart Contracts

This contract monitors the profiling of energy suppliers and the standing (reliability) of the energy consumers in the peers energy marketplaces. The profiling record of the energy product or service suppliers is associated with their prior execution in completing the energy trading, while the trustworthiness of the consumers rests within their energy consumption trace. The operation of the profile SCs is based on a profile account kept to each supplier, following the efficient conclusion of the energy exchanges. This account is revised after each trading with suppliers who complete their responsibilities in a committed way by obtaining greater score and thus, increased credibility in the energy marketplace. In a similar way, the reliability of the consumers is further screened by the profile SCs, with an attributed account reported on their record of energy consumption payments. The energy consumers who always compensate fully and promptly are expected to get greater reliability scores and thus, more prone to be entrusted by suppliers in the energy marketplaces. The contracts facilitate trustworthiness and liability in the peer-driven energy marketplaces, by offering a pattern for detecting the previous execution of energy partners. This activity tends to promote a legitimate energy contest and prevents suspicious actors from acting in an authorized way, thereby, promoting a rather more efficient and reliable-linked energy marketplace. The profile account of each energy supplier is transmitted as the total of all consumers scores offered, including their reliability recording. When the SCs energy suppliers offer incorrect rankings, they are expected to be associated with a lowered trusted validity profile, with their corresponding account revised and penalized, accordingly. In addition, once the energy trading organizational entity (i.e., the owner of the SCs platform) develops the running SCs and embeds the validated members (i.e., the decentralized energy consumers and their peers of decentralized energy suppliers), it offers the energy product or service producers with a preliminary credibility rating (i.e., advanced validity ranking). A summary outline of the above author's energy profile SCs shared verification infrastructure is showcased in the following Fig. 4.11.

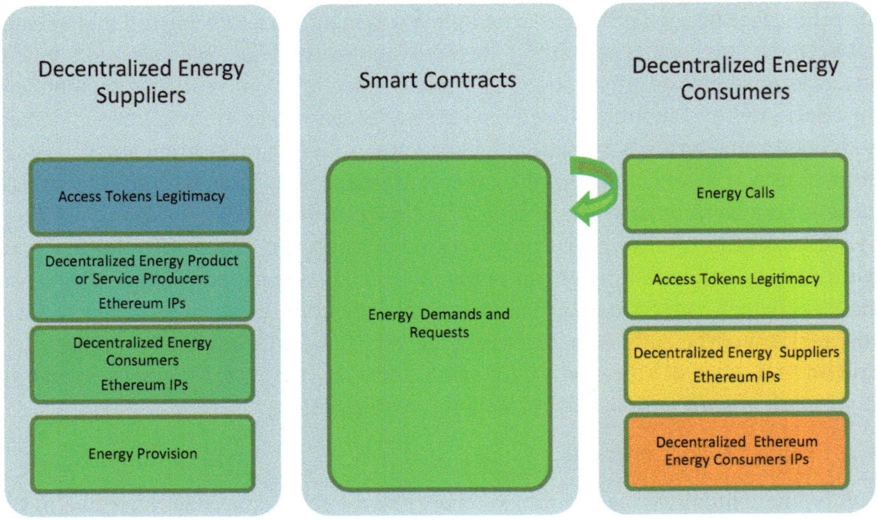

Fig. 4.11 The shared energy profile smart contracts authentication framework

Application and Assessment of the Novel Energy Smart Contracts Infrastructure

The envisioned authors suggested innovative framework is deployed through a decentralized implementation modelling using high-performance programming languages, with a twofold application protocol generated, and the energy SCs architecture implemented upon the Ethereum B ecosystem. Once the framework is deployed, it can be automatically run without any human-to-machine involvement. The effectiveness of the modelling was assessed in both public (open) and permission-granted (personal) B networks according to ecosystem's adaptability, authentication-driven running, and cost-management.

Adaptability-Handy Deployment

The above authors envisage that the agility of their framework can be examined by its capacity to manage modifications in the agents profile altered (e.g., number of energy suppliers and consumers differentiated, the profile energy SC record revised, etc.). For example, the amount of energy SC suppliers can be increased or decreased by taking away their Ethereum IPs (EIPs) from the SCs. New entries can be created by including newer EIPs without compromising the flexibility of the framework. In addition, the energy profile SCs are able to be further updated by revised SCs coding, re-implemented on the B ecosystem. In that respect, on their account, this infrastructure appears to garner greater adaptability across decentralized energy marketplaces. Additionally, the previous authors emphasize the benefits of revisions that can be performed across the one-to-one energy user (i.e., energy consumers and suppliers) trade network. Allowing, in that respect, greater scalable adaptations to differentiating energy marketplace status and (or) changes in energy customers and service (or product) suppliers attributes. Moreover, the capacity for an energy-driven trade enterprise to effortlessly connect to the peer-led energy exchange ecosystem seems to render further levels of adaptability, thus, permitting more facilitated testing with diverse B energy auction networks without the request for important assets allocated, and (or) ecosystem modifications. In essence, the visibility and permeability of the B infrastructure enables all energy trade revisions and moderations to the peer-to-peer auction framework being appropriately registered (documented) and confirmable by the engaged network-based agents (e.g., energy consumers and suppliers). This particular agency tends to advance the reliability and safety of the energy ecosystem, setting it as a rather "smarter" selection for energy trading businesses seeking to endorse and (or) be linked to an adaptable, safe, and visible one-to-one energy exchange platform.

Authentication-Based Safeguard

The modelling network bears a number of standards to guarantee that only validated agents (i.e., energy consumers and suppliers) can join the one-to-one trading platform. This agency can be accomplished through the implementation of energy SCs coding modifications. Such coding configurations can ensure that only the owners of the energy SCs (i.e., energy SC business) are able to perform particular operations such as embedding and (or) transferring decentralized energy consumers and (or) product or service suppliers, even modifying the standing of the peer-based exchange ecosystem. In case a non-legitimate energy agent tries to perform these operations, the performance will be unsuccessful. Likewise, other modifications are placed to guarantee that only validated energy agents can set energy offers, initiate access tokens (as above) and generate linkages. These modifications restrict accessibility to particular operations and make sure that only the prospective energy agents are eligible to join the peer-operated energy trading infrastructure. In sum, these norms are able to offer an empowering level of safety and guarantee that only verified energy agents can join the one-to-one energy exchange ecosystem, minimizing the danger of falsified and (or) illegitimate actions.

Fee-Based Benefits

The aforementioned authors claim that their framework infrastructure appears to be a rather cost-effective one. For example, their one-to-one energy-based trading platform contract fee is run just once to be established. In case an energy consumer and (or) supplier enters the ecosystem (i.e., either as decentralized energy consumers or decentralized energy product or service providers), the access-cost is minimal. In addition, when a consumer or supplier wishes to exit the network, the removal fee is equally the least. Further, disseminating an energy demand (call) on the Ethereum B, setting up an energy offer, finishing the energy exchange as supplier and (or) exiting an energy sale (auction) as consumer, requires the smallest payable amount. Even better, the energy profile SC operation under various ratings demands the lowest fee possible.

Safety Account

The fundamental Ethereum B ecosystem indicates an aforementioned author-based articulated profile of validated, flexible, visible and decentralized-driven secure mechanism. The robustness of the infrastructure is further supported by the energy trading recording that cannot be modified and (or) erased. The distributed-led mode of the adhered B platform safeguards that the network is of high-performance, scalable and adaptable. In addition, the authentication-based implementation structure tends to ensure that only the Ethereum-accredited (i.e., verified) energy consumers and (or) suppliers can join the peer-empowered energy trading platform and no reversed transactions can be issued once completed. Protection against malicious and (or) unsolicited attacks is empowered by advanced monitoring operation and robust energy SCs design and assessment. The energy SCs ecosystem was evaluated against potential vital threats. It appeared to be a resilient network, with no susceptibility towards core attack incidents, such as mistreatment of variables [e.g., "integer underflow" [27, p. 2], time marker reliance, Ethereum-based modification of trading cost and (or) re-access dependence. Additionally, the quantitative-led author assessment indicated robust ecosystem particulars. These properties were intertwined across the following tracks, as next. First, the capacity to manage vast amounts of data and clients without risking optimal execution (scalable-based). Second, the agency to be able to prevent illegitimate access and registration (safety). Third, the ease of user experience (UX) in terms of overall and energy SCs consumers and (or) suppliers-led, human–machine-interaction (HMI) [i.e., user interface (UI) and UX research]. Forth, the ability to adapt to modifying requests and (or) conditions (flexible-led). Fifth, the benefit to offer minimum fee and related value-added contribution (i.e., cost-efficiency). Sixth, the win to be less Metaverse-based computing power intensive, and more renewable resource eco-innovation infrastructure one. Seventh, the capability to ensure fair energy trading for all energy agents involved (e.g., energy consumers, suppliers, etc.). In total, the authors postulate that their fully-fledged, decentralized Ethereum B-driven energy ecosystem (with its sub-parts), appears to hold promising and rather robust performance indicators for a usable, optimized, trusted, safe, visible, flexible and cost-effective, eco-innovation energy trading infrastructure on Metaverse.

In a related domain within FinTech, elaborating on their work [24], the authors indicate that the B platforms in the energy context seem to continue to bear rather limited adoption and decreased energy marketplace prospects. In their current review, they report their interview-based findings garnered by leading energy sector suppliers and experts from information technology (IT) and FinTech businesses. This evidence regarded the technical, financial and regulatory challenges that appeared to hinder the inclusive and open mindset adoption (i.e., deployment and assessment) of the B platforms implementation across the energy Metaverse ecosystem. In a nutshell, the interview outcomes highlighted both the technical configuration and the crossover platforms antagonism as critical barriers in leveraging the B application across the Metaverse energy domain. The authors emphasize that the early developments of B platforms seemed to embed slower vending pace, elevated exchange cost, and sophisticated requirements from robust networks and fiscal marketplace directives.

In addition, the energy sector business models embedded were of limited innovative value, mostly restricted to transaction networks. Moving from the latter to the digital-first innovation systems and the appearance of the B Metaverse platforms as promising, service suppliers with improved energy transaction contract configuration might foster the transfer from the preliminary pilot use case systems to the disruptive, innovation energy commercialized ones. By endorsing energy consumers and suppliers in the prospective energy infrastructure, it is expected that energy marketplace ruling will lead to permitting one-to-one vendoring and decreased costs. The peer-to-peer energy trading prototyped on the Ethereum B seemed to be a continuing challenge in the later developed related work. It is only recently noted that additional challenges for energy trading ecosystems (i.e., including energy consumers and (or) service suppliers) have been revealed to be explored. The feature of the safety paradigm is a core one, indeed. The relationship between B and the SG seems to be investigated by an additional route of writing. This strand attends to the wins of distribution-based ledger technology when the energy network appears ever more spread out with an elevated percentage of irregular renewable energy endorsement from airstream or solar emission. The above authors project that some scholars have highlighted 4 probable user conditions of the B in the energy domain, namely, energy tokenized investments, one-to-one energy trading, electric car surge and electromobility. They also indicate that there seem to be critical hinders in adopting the B technological solutions in relation to their authorized and regulated status. In any case, though, they stress that none of the prior meta-analyses seem to offer a rather reasonable justification for the enduring breakdown of energy B tools to garner balance. The management of the energy ecosystem and the optimization of the infrastructure appear to be the competing forces behind the extended digital-first innovation energy network. The successful commercialization of the mechanism might not only rely on the platform's configuration, but also on the properties of the energy marketplaces intertwined. In describing their qualitative, structured interview data collection method, the author's insights direct towards exploring the barriers energy B platforms experience regarding extended commercialization (knowledge transfer) efforts. Their interview panel included a number of energy stakeholders, as next. First, a German energy business that deployed a B energy trading solution. Second, a global energy and B institution. Third, a UK-established Infrastructure-as-as-Service (IaaS) B designer, and a European Economic Area (EEA) energy investor going ahead. The preliminary interviews were supplemented by a control group that involved IT and FinTech representatives. The interview recurring themes corresponded to the prospective hinders of efficient energy knowledge transfer, reflecting those of technical viability, financial feasibility, implementation robustness, and that of regulatory capacity (i.e., validated, applicable, and to the point user cases) evidence. A brief outline of each of the four particulars that digital-first synergy B energy innovations bring to the table is described next.

Technical Viability

The above authors illustrate that due to their dispersed data deposit element configuration, Bs appear to bear elevated sophistication properties, yet consistent with their premise for technological advancements. Against the risk of attacks seeking to steal confidential data in a central storage infrastructure, the Bs are organized in a decentralized mode. This operation seems to ensure visibility, control and trustworthiness between network agents (e.g., an energy supplier that vends renewable energy and a business client that aims to utilize the "proof of origin" [55, p.3] to widely confirm that it adheres to particular energy-sustainability objectives. The influence of the B platforms tends to be assigned to their transparency and the unchangeable nature of their ledger-bas

ed transaction technology between users. The "Microgrid" energy domain enterprises denote a noteworthy example of the above eco-innovation and sustainable-driven autonomous and resilient via decentralization energy supply infrastructures. Such networks are largely run on regional renewable energy resources. In that respect, they constitute a critical asset of the next generation, eco-friendly energy ecosystem, by preventing energy breakouts in climate crisis events. The interviewees suggested that elevating pace and decreasing fees might work to sustain a "private or permissioned" (p. 14 above) B energy platform, permitting only chosen participants to trade energy. This approach might oppose the assertion of global accessibility, but would heighten effectiveness. The benefit of a public-based energy network could be partly counterbalanced by fears of data safety and confidentiality concerns, especially when considering the exchange between open and permissioned B energy platforms. This is an issue that has formed a universal discussion over the properties of Bs as global transaction energy ecosystems. The logistics sector tends to embrace a number of examples on successful B permission-based operation, with Ethereum, the largest B after Bitcoin, altering the verification mode to that of the proof-of-stake [6, p.7] in order to lower the energy expenditure. In this user case, the system is safeguarded by verifiers who must possess considerable shares of the corresponding energy cryptocurrencies. Other B instances also employ the PoS mechanism to decrease energy consumption. It seems, therefore, that the technical challenges of the early years of B platforms have been conquered, efficiently.

Integrated Energy Viability and User Occurrences

Resting on a systems design thinking approach (i.e., open innovation), an inclusive energy viability perspective tends to endorse a foundational agency across technical and knowledge transfer (commercial) viability, adhering to human-centred implementation. Firstly, when setting up B energy solutions, there needs to be a tangible and confirmed user situation on behalf of the enterprise that plans to employ energy-linked Bs. Following other interviewees comments by taking the example of an EEA energy industry, their B infrastructure established seems to decrease energy transaction fees. In a parallel run, a limited range of liquefied natural gas-powered was planned to be launched as a new business endeavour. However, these energy marketplaces seemed to be relatively restricted and complicated for a large energy stakeholder, as the interviewees organization. Therefore, it appeared evident that scalable-driven and generalization-led solutions were critical, and they perceived that as an actual user situation for the novel tools. In sum, the authors suggest that although Bs might reflect different business or digital-first innovation energy models, they appear to share their platform-based energy operation within and across platform-driven energy infrastructure(s). Tentative examples of these ecosystems are described below, as next.

Exchange Energy Solutions

A number of competing energy B platforms exist in current energy-led marketplaces. However, irrespective of possible financial gains (i.e., reduced fees), the B energy ecosystems tend to showcase challenging issues that resemble the challenges of other energy platforms, as well. They need to catch the attention of a vast number of energy marketplace members to accomplish favourable infrastructure exteriority, and preferably achieve a win-lose standing. For instance, there have been innovators who have invested in building a new energy network. However, they seemed not to succeed in attracting energy-traded securities from the current ones, because the maintenance impact of an energy marketplace which already bears extended transaction securities, tends to be advanced. Thus, they need to retain an increased financial expenses plan to lead securities to the new energy ecosystem, and this financial arrangement has to be kept far above the ground for long. On the opposite, a few B solutions in the energy domain mark rising energy marketplaces with no network taken over, for instance, in electric vehicle surge and (or) energy accreditation.

Innovation Ecosystems and Blockchain Infrastructure as an Energy Trading Service

Innovation energy networks function as the service software protocol for other energy trading solutions to be created by other industrial players. However, the majority of enterprises that operate innovation energy trading infrastructures also provide with the alternative to co-develop and (or) advocate other (external, outbound) energy suppliers with technical sophistication. This framework can also be perceived in the energy B platforms, as reported in the author's interviewee's responses above, aligning with other peers (e.g., [56]). Contrary to the energy trading platforms, innovation ones usually aim at diverse energy user indicators with various value added contribution. For instance, the associates of the energy service software protocol of interviewees are equally divided into three fundamental teams, as next such as energy-trade outsized industries, start-ups and investor (founder) businesses. Their motive to access the protocol also varies, as corroborated by interviewee's responses. In particular, for the energy business customers, it seems to be rather less expensive to enter an energy service software protocol, given that instead of doing research on their own, they are assigned to build upon an energy knowledge framework and swap practice. On the other hand, energy startup companies tend to be more interested in accessing prospective energy customers. The energy service software protocol also creates models that support energy-driven innovators to reduce the energy creation-driven production lines and the route to a least workable artefact. Further, energy platform users possess the value added contribution to become the energy assessors. They are able to obtain financial rewards when they operate an information processing system to support the energy infrastructure by data mining techniques. In certain cases, innovation-led energy ecosystems are antagonistic to private energy networks run and those service-handled by giant tech businesses. A number of enterprise-owned innovation energy ecosystems try to resemble their nonprofit energy peers by setting up clear and visible energy-fused governmental tactics. For instance, in the finance-energy domain, an interviewee commented that they use an energy network based on dispersed duties, lending tasks to a voted energy authority, and are supported through the energy ecosystem, which they possess. They are tasked to perform all their responsibilities by working towards confirming who's permitted to access the energy platform and also other business energy ecosystems. The teams that are expected to run the energy platform are going to define who will access their group(s). For the time being, therefore, the energy platform seems to run as an energy-led gatekeeper and (or) identification controller.

Akin to the energy service software protocol, innovation energy platforms operated by energy enterprise actors can be divided into two business units, with the first offering the energy-trade platform architecture and the other the energy-trade service. Some interviewees reported a user case from FinTech domain with an open source and a business version of the energy platform. They argued that both versions seem to bear the same nature of clients. The business one is employed by those planning to invest into business upgrading, often those entities prepared to move into production lines with the solution or those customers streamlining energy trading, because on all blockchain ecosystems the majority of the movement at present is tested. Most energy-fused activity is executed on "open source" [57, p. 1570, 1574], as the stakeholders seek to foster increased energy user activity and need energy clients to register for their business. The open source version is free for all energy users to utilize, while the enterprise one is licensed-fee based.

Contrary to the partnerships with energy-trading platform service, some other IT business interview representatives reported that they followed a service-driven enterprise model [58] resting upon the ecosystem energy utilities in order to support better energy customer services. This implies that they aim to work with diverse and extended range of teams from energy system controllers to logistics specialists, energy suppliers, energy manufacturers and new energy services brokers. The income generated on innovation platforms might adhere to diverse methods. The interviewees stated that their energy company-based model for the coordinating energy platform was a type of IaaS corresponding one. They operated this model in aid of the associates regulating and running it, under the license-based protocol. The partners remunerated them to oversee it, while they were further engaged in periodical energy client interaction. They also noted that the model for the energy marketplaces and (or) energy service artefacts (products) could vary between a payment per energy trading and a licensing one, respectively.

An alternative scheme some IT energy-trade enterprise players might follow would be not to enter any precarious energy ecosystem arrangements, but to provide interested business customers with energy solutions. For example, as indicated by some interviewee's reflective comments, it was evident that their business had not deployed a B platform at the energy production life-cycle, but they were ready to provide clients with blockchain-as-a-Service [55, p. 11] energy agency. They were involved with the regulation of popular open source, energy solution(s) service(s), in particular. By now, they were ready to handle three types of different B permission-based energy ecosystems. In addition, in a number of businesses, resting on clients requests, other B categories were also further deployed (e.g., healthcare). In sum, the authors outline that enterprises and startup businesses appear to rise above the barriers of energy platforms external (i.e., outbound) to internal (i.e., inbound) capabilities, by moving forward from energy artefact(s) and energy platform(s) company models with a high-performing regulatory agency to energy project management and (or) application enterprise models that deal with customized, tailored, personalized, not sophisticated and inclusive energy service provision.

Innovation Energy Trade Infrastructure, Engagement and Masking

Supplementary energy trading services and tools provision seems to be a critical factor in gaining energy marketplaces footing. Along these lines of practice, an intense competition to attract energy partners and especially those from diverse energy stakeholder teams is at place. According to some interviewee(s) comments, the way they perceive the energy partnerships network and that of their business, tends to align with three distinct energy agents. First, the current energy trading startups and (or) FinTech businesses that develop energy applications. Second, other IT-partner energy-trade players and (or) clients that wish to create an energy-based solution and prefer not to let someone use their resources, are expected to liaise with one of these energy partners to support the energy-driven solution development. Such associates are usually high-performance IT energy consultancy and (or) novel, B design energy companies, among others. Last but not least, the third energy agent could be energy trading platform (infrastructure) and (or) energy-trade cloud services, as well. In addition, it seems that autonomous innovation energy platforms perceive themselves as an attractive route for startup energy businesses. A few interviewees report that many startups approached them because of the public nature of their B platform, considered to trigger energy innovation. The nature of the business model for the energy platform's members seems to be a core issue. Indicative examples could include systems that reside within present energy applications, do not provide with masking, and developers (organizers) join via shares of the business enterprise and revenue, while members pay either subscription-based and (or) energy trading (exchange) fees. Masking tends to be a substitute alternative. Some interviewees indicated that their business was funded by energy token-based assets, at first. At the beginning, every associate had a least feasible fee for entrance. They were able to participate in the entire energy platform, and get energy token-driven tickets. At present, the tokens are withdrawn and the business has initiated a subscription payment for various energy customer categories. For startup energy companies, though, the membership is foundationally free of charge.

Energy Trade Regulation Models

The authors note that energy-trading B innovations seem to experience a two-fold, regulation-related challenging issue. On one side, electric power and natural gas networks constitute a section of the energy foundation infrastructure, with some aspects of the energy chain management serving as antagonistic agents, in specific, energy supply and wholesale, whereas, normally, grid-led energy trading supply and diffusion functions are managed and monitored by a state (i.e., government) bureau or committee. On the other side, Bs and other ledger-driven systems rest within finance regulations agency. When these regulation-based routes are combined, it is expected that an advanced level of sophistication for energy marketplaces' agents will be generated due to the two networks' interaction. For this reason, early developed Bs manifested multimodal hinders during their deployment in the electric power domain. For instance, product energy trading between agents on a B platform needs to adhere to specific regulatory monitoring and practices in the energy zone. An alternative to decrease the regulatory-oriented sophistication might be to separate the information material from the energy divisions and not to act as an energy-driven trading agency. However, the Bs are mostly connected to the transaction of data between energy marketplace customers. According to reported interviewees' responses, they argued that if they consider this regulatory line on a decentralized energy mode resting on Bs, this could be a mere route of communication that might be compared with other ones existed. They added that the novelty-based nature of the Bs has led a delay in the regulatory and legal-based energy marketplaces' feature identification, together with energy token-based management. They further emphasized the rather increased number of energy regulation-based barriers intertwined with the energy token-related activity, and the blurred horizon around legal responsibilities and accountability standardization for accreditors in the public-driven B. They noted that they could not see any context without any energy regulation-based challenges, especially when there seem to be no benchmarking on energy tokens and (or) money run returns. Some of the interviewees recommended that one-to-one energy trading might be facilitated by permitting multimodal energy service suppliers to provide intelligent sensing solutions (or devices). However, along this line of direction, the authors note that a critical enduring challenge would be both to sustain elastic associations and to confirm how and who is expected to claim responsibility across various grid operations. In addition, they argue that since new frontiers emerge on grid access and application(s), their open innovation-led modelling (i.e., systems design thinking, outbound to inbound networked exchange flow), tends to become rather more employable for the energy trading infrastructure services than for sustaining any multimodal planning already confirmed. In essence, they note that they opt to gain competitive, joint and unseen (energy) SCs. In any way, the (energy) consumer-service provider protection seems to be at the forefront of the EEA energy regulation authorities. The recent release of relevant cryptocurrency service codes of conduct by the European Council (EC) tends to denote this emerging trend. It seems imperative that universal digital energy asset regulation is critical. However, one has to reflect on the minimum regulation load needed to generate innovation (or open innovation) through energy customers' protection. Further, the authors emphasize that following up on cryptocurrency regulation paradigm, energy one seems to alter due to renewable smart energy trading solutions and resources generated. They stress that the EEA administration infrastructure promotes energy community-led customer engagement. Further, the urgency for energy safety due to the durable Ukraine emergency and escalating climate crisis events tend to provide with additional impetus for a global energy-led renovation. The environment-social-financial-community-and-government-connected drivers might be able to grasp, garner, diffuse and sustain visible, transparent, scalable, decreased energy and cost-effective B systems. In their interview(s) summary outcomes conclusions, the authors focus on the limited implementation that B energy-trade solutions bear since their generation, compared to the development of other B networks. Their three core arguments followed are as next. First, the antagonism between infrastructure energy platforms, with the preliminary ones experiencing combined technological, operational and competition-linked challenges. Owned to the public nature architecture of the earlier energy Bs, their related energy trading pace was restricted and their cost was advanced. Current energy marketplaces tend to seek high-performing authentication. Such high-standard authentication seems hard to provide with public Bs. The permission-based Bs, which are generated by enterprise stakeholders and (or) innovation-driven solution entities, such as the limited range energy trading ones of liquefied natural gas (LNGs [59, p. 1] receive important fee decrease. The second highest in profit, decentralized ledger technology-generated B, Ethereum, critically lowered its energy utilization, constituting it smarter for energy-trade consumers and (or) suppliers. Second, earlier B focus on offer(s) standardization and decreased procedure(s) integration, appeared to be rather deprived of seamless-driven energy consumer experience. Due to present and novel blockchain-as-a-Service (BaaS) and IaaS provision, energy artefacts (assets) and networks seem to move forward towards energy project management and application (system)-based corporate modelling, in a more personalized and embedded mode. Third, in terms of energy regulation-based sophistication, the energy-driven Bs ("mirroring decentralized ledger technology platforms) need to balance both core and finance-related infrastructure necessities. This agency tends to advance managerial density and decreases agent engagement. In the cases of commercially thriving energy NFTs Bs, these are digital-first energy innovations (open innovation ones) ruled by decentralized computational interaction, whereas one-to-one energy trading is managed not only through the networked energy platform, but also through the status of the regional grid (for instance, the provisional energy supply chain management roll out). The energy NFTs platform product minting tends not to get in the way of physical objects and (or) high technical requirements. On the contrary, if home providers of solar rooftops appear to vend excessive energy to their neighbourhood utilizing the present grid, in many provinces, they seem to be eligible as fully-fledged-linked energy vendors and thus, they need to follow different commercial-based regulatory conducts, as when joining regional energy marketplaces. The prior authors end up outlining that the most probable energy B innovation solutions to thrive are those meeting all three of the above conditions. In other words, operating in emerging-led energy-trade marketplaces, with value-added energy contribution and energy regulatory-based sophistication. The examples of B-validated accreditation might be seen as offering a high-performing authentication mechanism based on hard regulations compliance. These tend not to be hindered by high-performing regularity energy trading, as they can be vended as inclusive energy applications (systems) to enterprise stakeholders, provided by them in a carbon-limited energy regulatory scheme. In a nutshell, the commercialized and permission-based Bs might well serve the energy marketplaces, efficiently. However, such multimodal systems need to endorse robust authentication mechanisms supported by energy consumers, energy service providers and engaged energy stakeholders (e.g., proof-of-authority, proof-of-stake, etc.) [31, pp. 3,15].

Policy-Based Inferences and Next Generation Research

The aforementioned authors conclude that B implementation backed up by validation schemes seems to be still rather limited across the global energy-driven trading sector. Nevertheless, they recommend additional research to be headed for the following strands, as next. First, the development of PoS, permission-led B infrastructure across financial (and logistics) channels, jet-fuels, and electric-based surge. Second, the establishment of relevant regulation-driven paradigms that could serve the endorsement of energy consumers that trade energy assets in the universal energy marketplaces. In addition, the authors note the necessity to garner more quantitatively-driven energy data to highlight the current and future feasibility of the B implementation not only in the energy domain, but also across other sectors, worldwide. As a final remark, they outline that the pursuit for an energy trading B infrastructure supported by an environmentally-friendly (i.e., renewable resource-based) and regionally-associated energy supply chain management network, has been raised again. However, this challenging route along with that of how it could be potentially implemented on Metaverse ecosystem, are still ahead. Figure 4.12 depicts the inclusive summary of the authors B application and energy market-oriented commercialization challenges indicated, further expanded to integrate our tentative open innovation-verse (i.e., Metaverse) reflected operationalization, as follows.

4.3 Conclusion

The rarely envisioned future “EnergyMarketverse” trading centres are envisaged to revolutionize almost every dimension of smart intelligence and sensing-energy trading infrastructure. Built upon sustainable and eco-friendly, decentralized B-diverse, monetized, and smart technology-led energy product, service, consumer or supplier interrelated ecosystems on Metaverse, such markets are operationalized as bearing the potential to integrate immense smart computational intelligence, energy-efficiency and multiple user-centred, open-eco innovation energy trading properties. Despite their imagined, “EnergyMarketverse” next frontiers, distributed energy trading service efficiency, sustainability and open-eco innovation dynamics, yet, their projected current scarce inquiry is aligned with a multimodal nature of related wins and (or) challenges that hold the potential to be successfully addressed in the coming years.

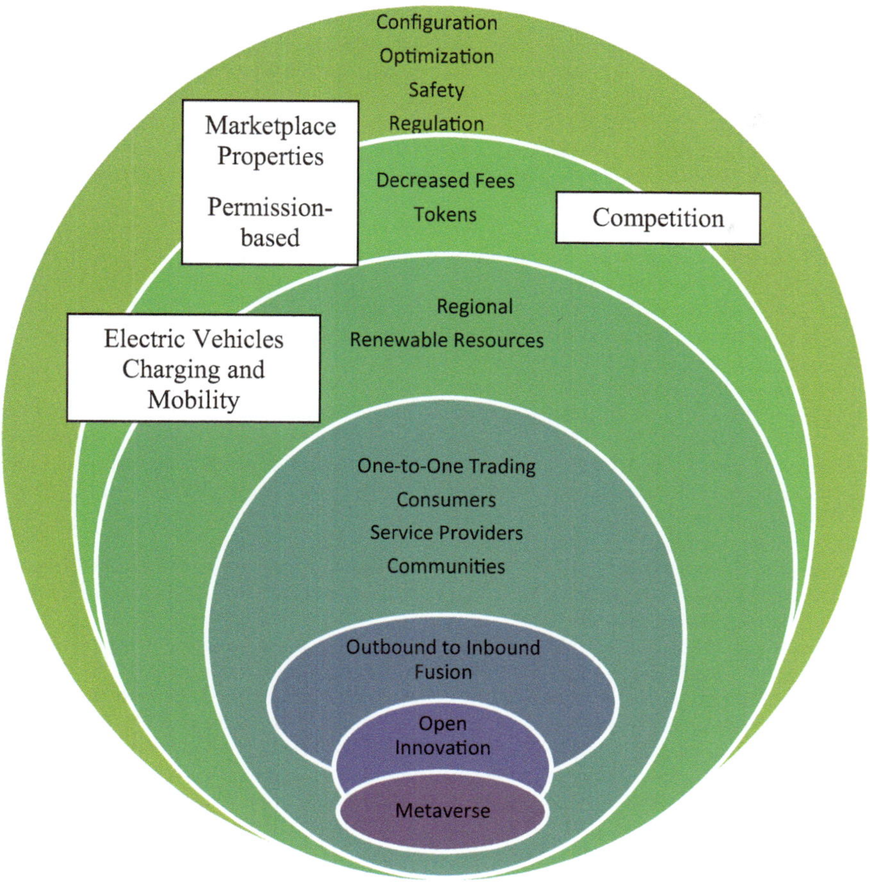

Fig. 4.12 Decentralized blockchain implementation and energy marketplace-commercialization open innovation-verse intertwined challenges

References

1. Asif R, Hassan SR (2023) Exploring the confluence of IoT and metaverse: future opportunities and challenges. IoT 4:412–429. https://doi.org/10.3390/iot4030018
2. Nleya SM, Velempini M (2024) Industrial metaverse: a comprehensive review, environmental impact, and challenges. Appl Sci 14:5736. https://doi.org/10.3390/app14135736
3. Liu S, Li P, Wang J, Liu P (2024) Toward industry 5.0: challenges and enablers of intelligent manufacturing technology implementation under the perspective of sustainability. Heliyon 10 (15): e35162. ISSN 2405-8440. https://doi.org/10.1016/j.heliyon.2024.e35162
4. Saridakis G, Khan Z, Knight G et al (2024) A look into the future: the impact of metaverse on traditional theories and thinking in international business. Manag Int Rev 64:597–632. https://doi.org/10.1007/s11575-024-00550-8
5. Nie Y, Hicks B, Nassehi A, Valero MR (2025) Aligning sustainable development goals and stakeholders' needs in food supply chains: a comprehensive review and mapping. Int J Prod Res 1–43. https://doi.org/10.1080/00207543.2025.260055

6. Bhujel S, Rahulamathavan Y (2022) A survey: security, transparency, and scalability issues of NFT's and its marketplaces. Sensors 22:8833. https://doi.org/10.3390/s22228833
7. Costa J, Moreira AC (2022) Public policies, open innovation ecosystems and innovation performance. Analysis of the impact of funding and regulations. J Open Innov Technol Market Complexity 8(4): 210. ISSN 2199-8531. https://doi.org/10.3390/joitmc8040210
8. Canseco-Lopez F, Serra A, Martorell Camps M (2025) Opening our innovation ecosystems to all: the INTEGER project case study. Sustainability 17:1164. https://doi.org/10.3390/su17031164
9. Llewellyn DWT, Paavo Ritala P (2025) "31: Open innovation ecosystems". Elgar encyclopedia of innovation management. Edward Elgar Publishing, Cheltenham, UK. https://doi.org/10.4337/9781035306459.00039
10. Oksana N, Kiseleva OV, Sysoeva, AV, Vasina V, Sysoev V (2022) Updating the open innovation concept based on ecosystem approach: regional aspects. J Open Innov Technol Market Complexity 8(2):2103. ISSN 2199-8531. https://doi.org/10.3390/joitmc8020103
11. Remneland WB, Styhre A (2023) Open innovation ecosystem organizing from a process view: a longitudinal study in the making of an innovation hub. R&D Manage 53:24–42. https://doi.org/10.1111/radm.12537
12. Huang J, Zhou P (2025) Open innovation and entrepreneurship: a review from the perspective of sustainable business models. Sustainability 17:939. https://doi.org/10.3390/su17030939
13. Costa J, Matias JCO (2020) Open innovation 4.0 as an enhancer of sustainable innovation ecosystems. Sustainability 12:8112. https://doi.org/10.3390/su12198112
14. Banga K, Gaile-Sarkane E (2025) Methods for measuring open innovation's impact on innovation ecosystems in the context of the european innovation scoreboard. Businesses 5:29. https://doi.org/10.3390/businesses5030029
15. Chesbrough H (2025) R&D management under disruption and uncertainty: an open innovation perspective. R&D Manage 55:998–1000. https://doi.org/10.1111/radm.12748
16. Cruz Romero RC (2025) As open as possible, but as closed as necessary: openness in innovation policy. Prometheus 40(3):166–181. https://doi.org/10.13169/Prometheus.40.3.0004
17. Alyami M, Javed MF, Ahmed WA, Haddad HA (2024) Examining the benefits, challenges, and drivers of open user innovation in small and medium-sized enterprises operating in low R&D industries. Heliyon 10(2):e24684. ISSN 2405-8440. https://doi.org/10.1016/j.heliyon.2024.e24684
18. Holgersson M, Dahlander L, Chesbrough H, Bogers MLAM (2024) Open innovation in the age of AI. Calif Manage Rev 67(1):5–20. https://doi.org/10.1177/00081256241279326
19. Heeren J, van de Vrande V, Volberda H, de Waard E (2024) Closing the innovation performance gap: open innovation in military bureaucracies. Calif Manage Rev 66(3):116–136. https://doi.org/10.1177/00081256241242166
20. Langa W, Mkwambisi D, Dougill A (2025) Systematic review of open innovation approaches for industrialisation in developing economies. Foresight STI Governance 19(3):17–33. https://doi.org/10.17323/fstig.2025.28979
21. Campos-Blázquez JR, Martín-García S, Cárdenas-Muñoz M (2024) Building an entrepreneurial ecosystem through Open Innovation fostered by public policies. J Innov Knowl 9(4):100587. ISSN 2444-569X. https://doi.org/10.1016/j.jik.2024.100587
22. Mais F, Bauernhansl T (2024) Design options for sustainable and open business models: a taxonomy-based analysis. Sustainability 16:4790. https://doi.org/10.3390/su16114790
23. Ma Z (2023) Energy metaverse: the conceptual framework with a review of the state-of-the-art methods and technologies. Energy Inform 6(42). https://doi.org/10.1186/s42162-023-00297-w
24. Burger C, Weinmann J (2022) Blockchain platforms in energy markets—a critical assessment. J Risk Financial Manage 15:516. https://doi.org/10.3390/jrfm15110516
25. Raman R, Kautish P, Siddiqui A, Siddiqui M, Nedungadi P (2025) The role of metaverse technologies in energy systems towards sustainable development goals. Energy Rep 13:4459–4476. ISSN 2352-4847. https://doi.org/10.1016/j.egyr.2025.04.012
26. Lezzi M, Del Vecchio V, Lazoi M (2024) Using blockchain technology for sustainability and secure data management in the energy industry: implications and future research directions. Sustainability 16:7949. https://doi.org/10.3390/su16187949

27. Naeini KH, Shomali R, Pishahang A, Hasanzadeh H, Asadi S, Gholizadeh Lonbar A (2025) PINN-DT: optimizing energy consumption in smart building using hybrid physics-informed neural networks and digital twin framework with blockchain security. Sensors 25:6242. https://doi.org/10.3390/s25196242
28. Wang X, Li H, Yi L, Ning Z, Tao X, Guo S, Zhang Y (2025) A survey on off-chain networks: frameworks, technologies, solutions and challenges. ACM Comput Surv 57(12):3131–3135. https://doi.org/10.1145/3735124
29. Jayavarma A, Parakkat Kesava Panikker P, Nair MG (2025) Revolutionizing the energy sector: exploring diversified blockchain platforms for a sustainable future. Front Blockchain 8:1544770. https://doi.org/10.3389/fbloc.2025.1544770
30. Batra P, Singh GR, Gandhi R (2023) NFT marketplace. arXiv:2304.10632v1 [cs.MM]. https://arxiv.org/abs/2304.10632
31. Barua S, Golder U, Chowdhury RS, Sharmeen K (2025) Implications of NFT as a sustainable fintech innovation for sustainable development and entrepreneurship. Sustain Technol Entrepreneurship 4(2):100099. ISSN 2773-0328. https://doi.org/10.1016/j.stae.2025.100099
32. Ren Y, Yang Z (2026) Open source innovation—a research review and prospects. In: Proceedings of the 2025 international conference on artificial intelligence and sustainable development (ICAISD '25). Association for Computing Machinery, New York, NY, USA, pp 397–404. https://doi.org/10.1145/3786484.3786545
33. Gustina A, Desty Nurmasari N, Chuan Liu JS (2024) Open innovation between university-industry: a review of research trends and practices. J Open Innov Technol Market Complexity 10(4):100419. ISSN 2199-8531. https://doi.org/10.1016/j.joitmc.2024.100419
34. Ming Ooi Y (2025) Is 2025 the year to rethink the impact of innovation activities and outcomes? Res Technol Manage 68(1):2–3. https://doi.org/10.1080/08956308.2025.2428136
35. Chesbrough H (2024) Open innovation: accomplishments and prospects for the next 20 years. Calif Manage Rev 67(1):164–180. https://doi.org/10.1177/00081256241273964
36. Osorno-Hinojosa R, Koria M, Ramírez-Vázquez DdC (2022) Open innovation with value co-creation from university-industry collaboration. J Open Innov Technol Mark Complex 8:32. https://doi.org/10.3390/joitmc8010032
37. Bogers M, Chesbrough H, Moedas C (2018) Open innovation: research, practices, and policies. Calif Manage Rev 60(2):5–16. https://doi.org/10.1177/0008125617745086
38. Strazzullo S, Mauriello R, Corvello V, Cricelli L, Grimaldi M (2025) How open innovation can improve companies' corporate social responsibility performance? Bus Ethics Environ Responsibility 34:1–16. https://doi.org/10.1111/beer.12535
39. Borges M, Kaminski PC, Guo J (rev ed) (2019) Differences in open innovation practices between headquarters and subsidiaries in the automotive industry: the French case. Cogent Eng 6:1. https://doi.org/10.1080/23311916.2019.1684806
40. Crupi A, Mortara L (2025) Building middle managers' managerial dynamic capabilities through collaborations: exploring the microfoundations in a university–industry consortium. J Technol Transf. https://doi.org/10.1007/s10961-025-10242-6
41. da Silva R, Kaminski P, Marin R (2021) Innovation ecosystems in the automotive industry between opportunities and limitations. Foresight STI Governance 15(3):66–80. https://doi.org/10.17323/2500-2597.2021.3.66.80
42. Machucho R, Ortiz D (2025) The impacts of artificial intelligence on business innovation: a comprehensive review of applications, organizational challenges, and ethical considerations. Systems 13:264. https://doi.org/10.3390/systems13040264
43. Broekhuizen T, Dekker H, de Faria P, Firk S, Nguyen DK, Sofka W (2023) AI for managing open innovation: opportunities, challenges, and a research agenda. J Bus Res 167:114196. ISSN 0148-2963. https://doi.org/10.1016/j.jbusres.2023.114196
44. Hassan MU, Rehmani MH, Chen J (2026) VPT: privacy preserving energy trading and block mining mechanism for blockchain based virtual power plants. ACM Trans Priv Secur 29(1):1–35. https://doi.org/10.1145/3767163
45. Tightiz L, Minh Dang L, Padmanaban S, Hur K (2024) Metaverse-driven smart grid architecture. Energy Rep 12:2014–2025. ISSN 2352-4847. https://doi.org/10.1016/j.egyr.2024.08.027

46. Lee CC, Yeh WC, Yu Z et al (2023) Knowledge sharing and innovation performance: a case study on the impact of organizational culture, structural capital, human resource management practices, and relational capital of real estate agents. Humanit Soc Sci Commun 10:707. https://doi.org/10.1057/s41599-023-02185-w
47. Rumanti A, Rizana FR, Reynaldo R (2021) The impact of open innovation preparation on organizational performance: a systematic literature review. IEEE Access 9:126952–126966. https://doi.org/10.1109/ACCESS.2021.3111091
48. Iqbal S, Bashir M, Bureš V, Zanker M, Ullah S, Rizwan A (2025) Impact of knowledge sharing and innovation on performance of organization: evidence from the pharmaceutical industry of Pakistan. Cogent Bus Manage 12:1. https://doi.org/10.1080/23311975.2025.2471536
49. Ngo QH (2023) The effectiveness of strategic alignment between open innovation and generic strategies: empirical evidence from restaurant SMEs in Vietnam. J Open Innov Technol Market Complexity 9(1):100016. ISSN 2199-8531. https://doi.org/10.1016/j.joitmc.2023.100016
50. Portuguez-Castro M (2023) Exploring the potential of open innovation for co-creation in entrepreneurship: a systematic literature review. Adm Sci 13:198. https://doi.org/10.3390/admsci13090198
51. Audretsch BD, Belitski M (2023) The limits to open innovation and its impact on innovation performance. Technovation 119:102519. ISSN 0166-4972. https://doi.org/10.1016/j.technovation.2022.102519
52. Wu Q, Sánchez-García E, Marco-Lajara B et al (2026) Unveiling research fronts in knowledge management for innovation and sustainable development. Environ Dev Sustain. https://doi.org/10.1007/s10668-025-07261-6
53. Băban M, Băban CF (2025) Patterns of open innovation between industry and university: a fuzzy cluster analysis based on the antecedents of their collaboration. Mathematics 13:772. https://doi.org/10.3390/math13050772
54. Mele C, Russo Spena T, Russo SP (2025) Experimenting on the metaverse to foster innovation. Spanish J Marketing—ESIC 29(3):270–292. https://doi.org/10.1108/SJME-05-2023-0117
55. Schmid J, Ubacht J, van Engelenburg S, van Oudheusden J, Chappin E (2024) Is it green? Designing a blockchain-based certification system for the EU hydrogen market. Front Blockchain 7:1408743. https://doi.org/10.3389/fbloc.2024.1408743
56. Santos L, Gomes A, Rupino P (2025) Energy trading using blockchain: smart contracts functionalities—a systematic review. Energy Strategy Rev 61:101825. ISSN 2211-467X. https://doi.org/10.1016/j.esr.2025.101825
57. Meng H, Ding JH, Wang Z, Zhang XY, Ning H (2025) Blockchain enabled metaverse: development and applications. Tsinghua Sci Technol 30(4):1552–1582. https://doi.org/10.26599/TST.2024.9010054
58. Zhao H, Zhao J, Liu W, Yan Y, Huang J, Wen F (2023) Metaverse-based decentralised autonomous organisation in energy systems. Energy Convers Econ 4:379–386. https://doi.org/10.1049/enc2.12104
59. Kemfert C, Präger F, Hoffart FM, von Hirschhausen C (2025) Locked in a fossil-centric system paradigm: LNG expansion impedes socio-ecological transition toward a just and renewable energy future. Cell Rep Sustain 2(8):100464. ISSN 2949-7906. https://doi.org/10.1016/j.crsus.2025.100464

Chapter 5
Campusverse and Industry-Smart Grid Energy Learning and Development

Abstract The Metaverse might be operationalized as a next generation open innovation-led learning and development (L&D) ecosystem both within higher education (Smart Campusverse) and across business and public contexts that can support academics, researchers, consumers, energy providers and government with Smart Grid sustainable energy L&D. The following chapter attempts to outline the sparse yet developing research evidence on the emerging Metaverse-based innovations that could facilitate immersive instruction and meaningful learning for the Smart Grid(s) renewable resource L&D in academic and business entities [K-Cube Virtual Reality (VR) systems, Energy Battery Mixed Reality (MR) Learning Games, Smart Factories]. Corroborating to the above, the chapter further aims to sketch challenging issues associated with the future use of Metaverse-fused emergent solutions for Smart Grid(s) sustainable power training, evident in education and industry settings alike.

Keywords Campusverse · Smart grid energy · Education · Learning and development · Energy battery · Smart factory

5.1 Introduction

Tan et al. [1] stress the progressive development of gradually matured emerging technologies and in particular, a number of Metaverse-led ones [e.g., augmented reality (AR)/VR systems]. However, their research exploring AR/VR solutions implementation in education (higher education) and (or) business for engineering-based L&D (i.e., safety training, accident prevention, etc.), revealed limited, yet developing usage. Their meta-analysis indicated that 59% of the generic AR/VR tools are used to instruct learners and (or) employees in operating intricate equipment (or devices), with 41% of them employed to instruct students and (or) workforce how to detect risks and (or) avoid accidents, efficiently. Within engineering sector, in particular, 16.59% of AR/VR systems are adopted for AR/VR-based operation manual and 11.41% are used for AR/VR-associated safety training, respectively. The AR/VR tools are slowly but surely offering the chance to transform education (tertiary education)

E. G. Makri, *Metaverse and Smart Grid in Energy Industry: An Open Innovation Challenge on Trends and Applications*, Studies in Smart Technologies,
https://doi.org/10.1007/978-3-032-29666-5_5

and business L&D (training), by providing with "real-life-and real-time" replicated immersive instruction on challenging accessible domains. The authors conclude that improvements in heads-ups displays (HDs) instructional design for greater user experience, increased cost-effectiveness and additional integration of advanced cloud computing and mobile networks technology for better human–machine interaction, are indicative examples of AR/VR future usability and implementation challenges to be addressed.

Yaqoob et al. [2] indicate that future Metaverse seems to be promising for the energy zone, particularly in adopting renewable power resources. In terms of potential or expected benefits for the energy industrial domain per se, the authors emphasize the following reported advantages, as next. First, the Metaverse's possible capacity of multiple way interaction channels between the energy providers and the end users that can facilitate virtual customer support and branding. Second, the expected provision of a shielded and an attractive mixed reality space for L&D (T&L, professional practice) by using the digital twin (DT) technology to assume real-world and real-time conditions. Third, the Metaverse might well support the design (or co-design) of manufacturing replications for assessing energy ecosystem modifications, therefore corroborating to expenditure decrease, less faults and evolved discharges assessments. Embedding the Internet of Things (IoT) technology and Artificial Intelligence (AI) with VR equipment offers an option to acquire more accurate awareness and practice of power utilization and cost-effectiveness. Diverse solutions like AI/VR/AR/MR/IoT/3Dimension(D) re-modelling and Blockchain (B), may be adopted to empower the energy and utility branch on Metaverse platform. These emerging and cutting-edge technologies are able to support distant upkeeping and control of resources, VR space drop-ins, workforce training (L&D), real-time resource control and diaphanous and effective power auditing. In addition, the utilization of these solutions may permit energy businesses to extend their facilities to customers and promote smarter groups as delivery utilities administrators. Effective energy use may be accomplished by implementing the Metaverse within the Smart Grid (SG). For example, the Metaverse seems to provide with the option to design (co-design) VR energy factories consisted of dispersed power resources (e.g., airstream turbines, solar panels and battery-operated repository machines), all handled by digital systems. The energy oversight on Metaverse platform may be considerably advanced through replication and upgrading. Energy customers and suppliers might be able to inform better decision-making by envisaging power usage in real-time mode. For instance, envisioning energy adoption schemes through VR user interface may drive into reduced energy use in high demand times, with VR energy factories and actual screening of energy consumption being the key drivers of power management advancement.

Sin et al. [3] report that the Metaverse is becoming an attractive notion and subject matter for visionary academic, research and business immersive innovation in recent years. Corroborating to their argument, we suggest that beyond discussed topics on Metaverse's security, monitoring, maintenance and energy management, the future Metaverse for either education (all levels) and (or) L&D (industry training) seems to denote noted examples of related inspiring topics. In particular, the questions of what a Metaverse for higher education ("Campusverse") or Metaverse for L&D (professional practice) ("Learningverse") should look like?

Addressing the need to expand upon limited research on Campusverse course administration, interventions and assessments and the one that connects Extended Reality (XR) (Metaverse) and Knowledge Graph instructional learning design management (e.g., reflecting mind mapping visualization), the authors elaborate on their Metaverse-linked Knowledge graph (K)-Cube VR system, co-designed (i.e., open innovation) to engage adult learners in Campusverse instruction and reflect upon their designated user experience learning outcomes. Following recent debates, they operationalize Metaverse as a digital space that offers admission to networked subject topics through physical replications that can supplement tasks or events more challenging ones to accomplish in actual life. They perceive Metaverse as a universal and ceaseless digital ecosystem with accessible contents similar to Internet service provider. This platform appears to be attractive via interaction(s) that rest(s) upon physical replicas (i.e., replication of the actual world) and targets at integrating diverse events that range from commercial to industrial and educational ones. Reflecting upon their own Metaverse K-Cube VR solution, the authors argue that it relates to recent pedagogy closely exercised upon or applicable at the Metaverse platform. Collaborative virtual space learning, immersive L&D and visualization pedagogical features are coupled with reflection upon own practice and conceptualization of the world (constructivist learning) to guide or capture the learners' subject matter provision and their user interface experience. The above educational trends implemented at the prototype are structured in the mode of a knowledge graph (p. 5) with networked educational notions, envisioned as investigable virtual reality environment. This VR setting is also expected to have a multimedia hub, multiple user avatar interaction, an open space that connects learning chunks and facilitates the design (or co-design) of the learning content and additional personalized learning and social presence elements (i.e., "sense of being there"). The ultimate goal is to inspire students to explore in their own way an interrelated array of internal and external aspects of learning content creation through a shared immersive (3D) platform. The authors illustrate their pilot user experience case study on their K-Cube VR system. The participants were required to use the prototype to comprehend a synopsis of a course in a self-directed way. The awareness of the curriculum is divided into index terms and their relations, which are inspired in the format of chunks and margins, accordingly. The attendees were also requested to prepare for and deliver an outline of a course of study by acting on a 3D knowledge graph display and looking at files on the screen. The authors performed focus group interactions to garner the learners' insights and feedback on their experience with the prototype. Their overall conclusion was that the Metaverse-led K-Cube VR tool appeared to improve the participants' comprehension of the curriculum. This favourable learning outcome is attributed to the potential engaging integration of the knowledge graph's instructional design within the VR immersive delivery (transferable, experiential and absorbed learning through VR system), corroborating to the robust learning potential of the "future" Campusverse.

The authors elaborate on their Campusverse-led framework introduced as the expected to be a fully-developed solution at their K-Cube VR prototype, as next. Their solution is showcased as a knowledge graph (KG) (p. 6) where a (learning) chunk (i.e., node, p. as above) depicts a knowledge (learning) index term. The participant is able to pass through from one (learning) tag to the other, each of which tends to act as a multimodal hub for looking over the learning space. Avatars and visual cues support the contact concerning the learning material and the learners' screening. The knowledge graph allows the appropriate management of and access at the immersive learning context. The latter can be extended within the Campusverse learning space. Additional virtual social interaction features can ensure real-time responses and comments that allow for learning material suggestions. Further particulars of their K-Cube VR system are as follows. First, their knowledge graph is expected not to virtually replicate the real world per se in the form, for example, of a DT (often seen on Metaverse), but the knowledge per se that relates learning notions and context. Replicating an Internet service provider, the knowledge graph resembles a web link that connects affiliated background and offers swift access to interrelated information. In that way, the participant is able to view and reflect not only on how data, information and (or) learning context is constructed (constructivist learning), but also on how prior knowledge is used to build newer one by visualizing connected concepts concurrently. This feature intends to foster the learners' exposure to newer knowledge and (or) notions, making them explore diverse constructs in an open, networked mode simultaneously. The authors expect that their knowledge graph can support curricula across different subject domains. Further, the integrated avatars are able to support the learners' showcasing by drafting the hands to the users' tracking controllers or real hands. Such "hands" can assist the learners to showcase at the learning material in the Campusverse and share their views with their peers. In addition, the avatars' vision can mirror where the participants are looking at to assist other attendees realize what the learners are concentrating on. The authors' proposed K-Cube VR solution is expected to be applied as an open, Campusverse accessible system where learners will be able to use a customized immersive learning environment in an inflow and outflow networked (i.e., open innovation) visualized interaction. Corroborating to this open-ended multidirectional virtual learning creation space, the researchers embed two educational subject design tools connected to the knowledge graph: that of the logical reasoning to support the one of item emplacement (i.e., out-VR for tutors and in-VR for learners) (p. 6). Lastly, they finalize their system by incorporating social channels elements and connect them to index terms (or tags) to facilitate the learners' sharing of their insights and (or) suggestions for additional meaningful learning content(s). Figures 5.1, 5.2 and 5.3 replicate the authors' open, K-Cube VR Campusverse system integrated components.

Their K-Cube VR present status prototype integrates the features of knowledge graph Campusverse presented above (Figs. 5.1, 5.2 and 5.3), incorporating the curriculum overview (introductory course summary) as the first case study explored for this solution (important application on Campuverse). This preliminary curriculum introduction deems to be an important option to support university learners choose the most suitable courses for them in a self-directed learning delivery. By using the

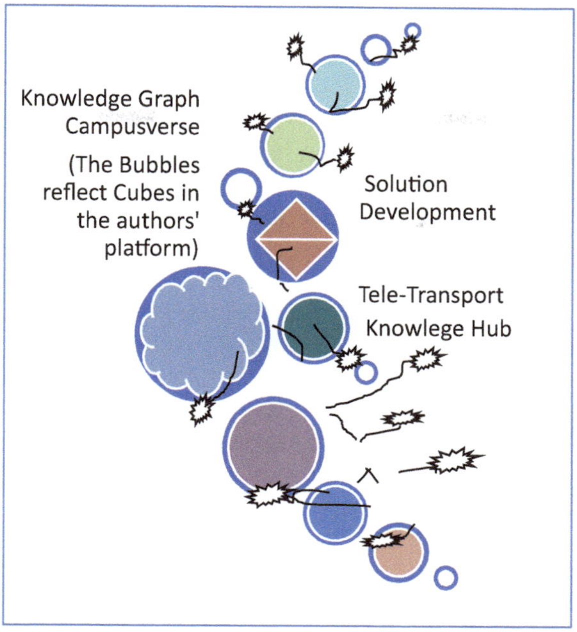

Fig. 5.1 The K-Cube VR platform with social learning

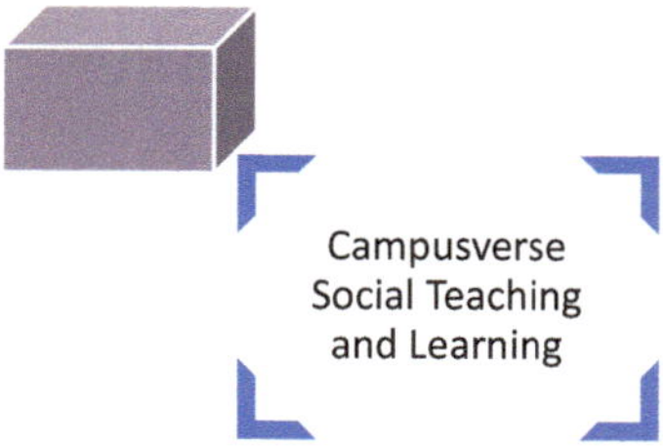

Fig. 5.2 The K-Cube VR platform with social learning

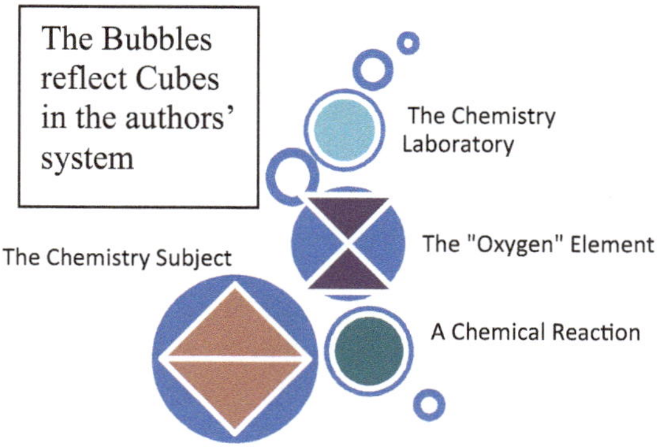

Fig. 5.3 The K-Cube VR prototype indicative example course overview

K-Cube VR current version, the learners are expected to view an array of index terms (keywords) through a head-mounted display (HMD). A cabin index term depicts the curriculum while each stellar keyword denotes the tag connected to the related tutorial learning material and Wikipedia site (p. 8). When the participants showcase at an index term, the tutorial learning slides will arise. At this point, the learners are able to use a joystick to navigate through the left and right sides of the tutorial. To see the Wikipedia site, the students can act on learning keys presented across the learning content. The latter is expected to vanish when the participants are not acting on its associated index term. The users can also hitch keywords by clicking on the keys. When a keyword is hitched, it can be grasped to a preferred place for more flexible observing. In order to position a keyword at a particular point, the participant can first hitch the keyword and then he (she) can grasp it by pushing down the index and hand activators. The attendee can then shift the keyword, while grasping, to the new place. In addition, for the two dashboard-based material (tutorial slides and Wikipedia site), the learner can adjust the balance of the dashboard to assist with his (her) learning material overview.

In order to assess the user experience of their co-created (i.e., open innovation) K-Cube VR prototype introductory course for Campusverse, the authors elaborate on their pilot study performed to garner prospective focus groups insights and feedback from learners (gamers) regarding their expected knowledge acquired and tasks executed during the multimodal K-Cube VR practice. With a view to explore the two modalities tools integrated (i.e., VR and KG used on desktop), a third one, that is, a conventional course introduction mode was employed as control group. In that respect, the participants were introduced to a standard format of files on the left and Wikipedia on the right hand sides, respectively. The attendees were also able to view the KG desktop modality slides on the left. Using the mouse, the learners could navigate around the KG interface moving across the tutorial material by selecting the corresponding notions each time on the left. 45 university students served as the engaged participants through prospective advertisements and classroom groups. They were assigned to explore the multimodal K-Cube VR solution by forming each modality team with 15 users, at random. The control group consisted of 3 adult women and 12 men, their desktop peers included 5 females and 10 males and their K-Cube VR system engaged 3 women and 12 men, accordingly. 16 of the participants reported prior practice with VR tools, 5 of which assumed the control group, 6 their desktop peers and 5 their K-Cube VR members, respectively. The authors prompted the learners to navigate around the tools as much as possible, without any hard timeline, in order for them to be able to gain a better outline of the system. The users can navigate the application, that is, the K-Cube VR by tele-transport, chunk (i.e., node, p. 9) tele-transport, with each index term associated with a tutorial slide material and a Wikipedia web page. The students can move the chunks (i.e., nodes) by grasping. A circle (or cube) denotes a curriculum and a star represents a learning keyword.

The authors indicate that the devices used in their experiment for the K-Cube VR were an Oculus Quest 2 HMD and a right-handed controller (p. as above). The control and desktop modes were delivered through a screen, a keyboard and a mouse, respectively. They first introduced the different system modalities to the attendees, by showcasing how they could navigate their way across each one of them in a detailed manner. Then, they administered the user experience questionnaires [i.e., knowledge acquisition per course and workload assumed to execute the modality application (NASA Task Load Index, p. 10)]. Next, they performed the focus groups discussions in 3 to 4 sessions [with topics including the range of understanding, degree of immersion experienced, the depth of learning concepts covered and the technical (digital) challenges met when serving each of the system's modality] to garner the attendees' emotions and perceptions about the course content and the presenting mode. The participants identified 30 index terms (keywords) with 7 associated with the curriculum. In terms of performance indicators, the K-Cube VR users showcased the best score, followed by their Desktop peers and the Control group ones, accordingly. The VR participants were indicated as the least keen on choosing un-associated index terms (keywords), after their Desktop and Control group members, respectively. The course's learning engagement and users' perceived confidence seemed to be greater among the K-Cube VR modality, compared to Desktop and Control one, accordingly. In other words, it seemed to rather validate the expected meaningful learning practice and visualization immersion that accompanies the Campusverse platform. In addition, the difference between the three modalities (K-Cube VR, Desktop and Control group) in terms of workload assignment appeared to be non-significant, therefore, it seemed that the immersive VR-Campusverse modality did not appear to impose any additional workload to the users compared to the other modes. The focus groups feedback indicated that the structure of the curriculum was the meaningful learning contribution for the Control group users, that of the usefulness of the course for their Desktop peers and knowledge visualization that supports learning memory, interrelated mind mapping and easy access for the K-Cube VR users, accordingly. A number of K-Cube VR learners reported that they sensed the feeling of social presence in the virtual modality, felt more focused and absorbed by the immersive learning delivery (i.e., multimodal mode) that supported them not to keep track of time interval during interaction. Moreover, the users perceived the Campusverse platform as one that facilitated their learning effort to build a self-directed inflow mind map that tends to foster the learning abstraction and comprehension, effectively. They further viewed the VR user interface as scaffolding one, of concrete patterns and with decreased manual adaptation that foster greater ease of user navigation. Some of the users reported some degree of fatigue due to heavy HMD device, corroborating to the application of lighter AR equipment to XR adaptable solutions. Overall, the authors outline the promising learning outcomes that their K-Cube VR Campusverse space revealed for higher education students, partly confirming the meaningful Campusverse platform immersive teaching and learning.

Hare and Tang [4] indicate that Metaverse seems to offer a great space for digital-physical-social user interfaces that can evolve L&D in a personalized and immersive machine-human interaction. However, they emphasize that despite the generic interest in Metaverse parts and gains, there seems to be limited research on the technological advancements behind effective Metaverse teaching and learning. In particular, they argue that there tends to be lack of research not only on how to track the learners' actions on Metaverse ecosystem, but also how to ensure attendees' dynamic adjustment and provision of relative support during interaction. In their current work, the authors propose a technological approach (model) to overcome this reported breach. This framework is built around a hierarchical, multiagent and shared reinforcement learning modelling (p. 2047) that aims to promote multifaceted, adaptable and personalized Metaverse learning provision. The structure (co-design, open innovation) and the advantages of the suggested approach are showcased in a Metaverse learning game (i.e., "Gridlock") followed by expanded replicas. More specifically, their proposed framework aims to create a digital-physical-social ecosystem that employs personalized computer–human interaction to foster teaching and learning with actual (real life) human learners and personal computer (PC)-monitored AI-non-player agents. The latter are assumed to take 3 types (skilled, apprentice and tutor: p. 2048) that can empower shared, reflective and evolving learning interaction. They are able to screen learning outcomes by getting feedback on learning challenges and performance, to garner data analytics to be able to predict future learning performance and decide on best routes available to support meaningful learning. The above technical features were embedded into the co-design (i.e., open innovation) of a Metaverse learning game ("Gridlock"). The game intended to support a computer engineering curriculum attendee collection data. This was made possible through experiments that were built upon the data to validate the effectiveness of learning behaviour exploration. In Gridlock, the learners assumed the role of an engineer that was tasked to design a traffic light controller. Under the authors-led approach, the participants play the tool in the physical and virtual environment. The solution divides the traffic light design into eight themes, each of which assigns attendees with context-specific tasks. An advanced rank agent and eight lowered rank agents are used to create the AI-non-player agents that leverage the participants' learning insights, get their learning performance indicators and personalize their instructional practice. The agents' tasks and behaviour are transferred to as learners' support. The advanced rank agent collects feedback across different virtual space sections, each one corresponding to diverse learning game subject(s). Next, the attendees are directed towards the sub-subject areas themes where the eight lowered rank agents are tasked to run. During certain game time intervals, these eight agents are allowed to get support from expert-led offered collection. The participants' performance indicators consist of assignment scoring, time needed to complete the assessments and navigating through learning game levels, emotion tracking and complementary index of learners' gaming record. The lowered rank agents are programmed to select functions that improve the attendees' assessed behaviour. Similarly, the advanced rank agent gets rewards for elevated performance in subject areas where the participants seem to indicate decreased achievement. All rank-led agents are tasked to offer as little learning support as possible to the attendees during the game. In order to validate various features of their technical framework, the authors performed two computer-led experiments. Their experiments aimed to garner comparative evidence across different advanced and lowered rank agent-based conditions. Their first experiment integrated the comparison of the advanced rank agent with the eight lowered rank ones, with the learner interacting first with the solution and then receiving support in all learning actions until their performance indicators were successful. The authors reported that their hierarchical methodology taken appeared to offer concrete evolving standard reward during preliminary instruction. The second experimental setting included only the lowered rank agents that offered task-particular support. Four agent conditions were selected, as next. The first included one agent, the second multiple agents, the third multiple agents with shared practice (i.e., reinforcement learning) and the fourth one with the authors' suggested balanced shared exercise. The latter case (i.e., the authors' proposed approach) surpassed the rest ones across the rewards' scores. The researchers appear to confirm that all the multiple agent cases showcased improved achievement than the one-agent condition. In that respect, the authors seem to validate the use of multiple agent approach to promote and sustain more context-specific learning behaviour, whilst the one-agent framework would demand more experience from the agent to assimilate and perform generic learning knowledge and skills for all actions. In addition, they outline that without the acquisition of balanced knowledge from various agents, the solution tends to achieve inferior to an average multiple agent framework. This piece of evidence appears to corroborate to their premise that mutual exercise needs to be balanced as it does not seem to completely reflect the potential of the intended agent. In summary, the researchers appear to offer a cross-functional AI-led, non-player agent co-design (i.e., open innovation) powered with reinforcement learning approach to trigger personalized Metaverse education, as efficiently implemented on Gridlock learning game. Their experimental assessments revealed that both the hierarchical multiple agent condition and shared practice cases lend support to agent learning effectiveness and decrease in metrics constraints across Metaverse learning. They conclude that the implementation of more computational experiments extending their current limited usage is expected to shed further meaningful insights on optimal configuration of Metaverse instructional ecosystem.

Given the developing global energy demands, Janev et al. [5–7] introduce the role of smart (i.e., intelligent, cutting-edge) technology in renewable power generation, diffusion and storage across a rather undemanding but efficient ecosystem. Reflecting on their insights, we agree that the Smart Grid (Power) Technology (SGT) seems to be able to provide us with this exclusive element, in empowering the full potential of power production and deployment in a networked delivery, as operationalized in Fig. 5.4.

Aligning with the above authors rationale and reasoning, we endorse the SG as a notion that amalgamates a fully-fledged energy (power) interrelated generation and transmission ecosystem in an integrated (inclusive) infrastructure. As inferred from its broader conceptualization, the Grid (Network) is Smart (i.e., intelligent, novel), that is, an ecosystem that aims to provide us with clean power, as mandated in recent years at a global rate. Supporting the aforementioned author's insights, we agree that the Smart Grid Power Compound (SGPC) is considered to offer an adaptable, quality, monitored and secure clean energy infrastructure that tends to be self-controlled and counteractive through a circulated computer-aided system. In addition, we identify with their narrative that "echoes" the Conventional Grid (Energy) Infrastructure (CGEI) as rather slower-moving, with questionable robust

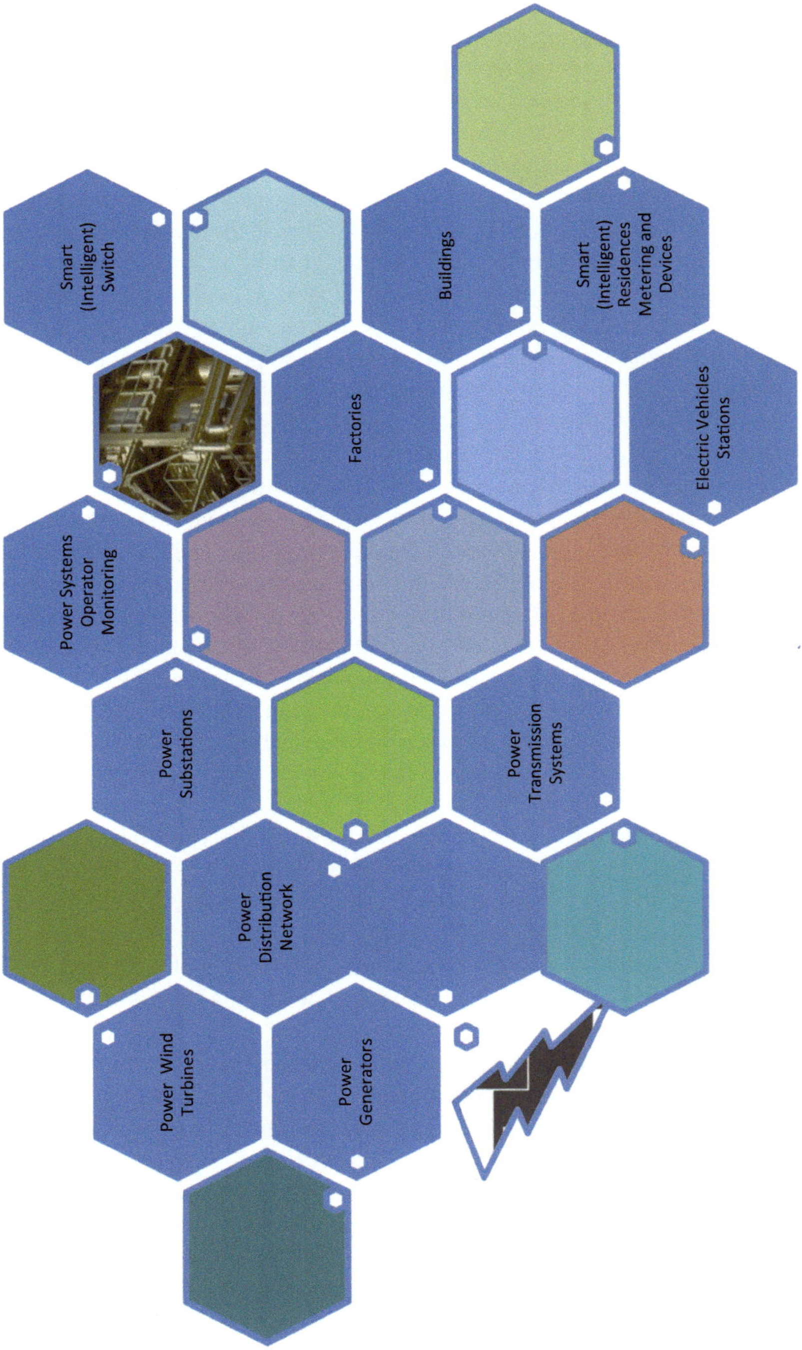

Fig. 5.4 An envisioned illustration of the smart (intelligent) grid infrastructure

end user security and flow of real timeline and continuous data transmission in-between the power suppliers and clients (consumers), across any stage of energy generation and delivery (see Fig. 5.5 next) (Fig. 5.6).

Further reflecting on their rationale, they align with the premises that the rather weak arrangement of power measurement seems to make the Grid open to energy leak and loss across regions. In addition, the challenge to locate any energy fault in the ecosystem is made rather not possible to be achieved, due to restricted safety infrastructure. In cases of energy leakage, therefore, the whole grid might shut down until the error is located and fixed. On the other hand, the Smart (Intelligent) Grid

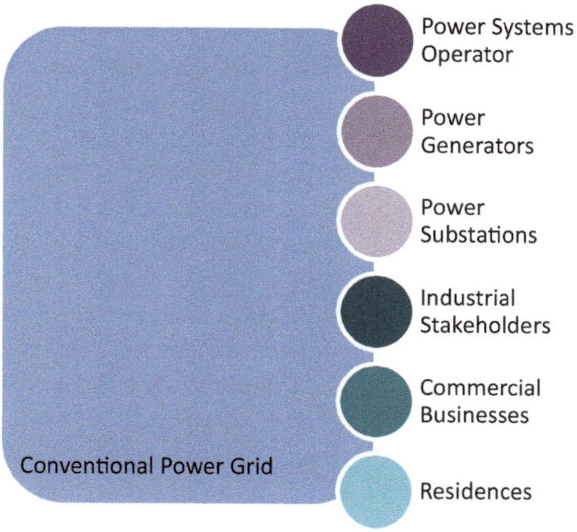

Fig. 5.5 Reflective depiction of the conventional grid

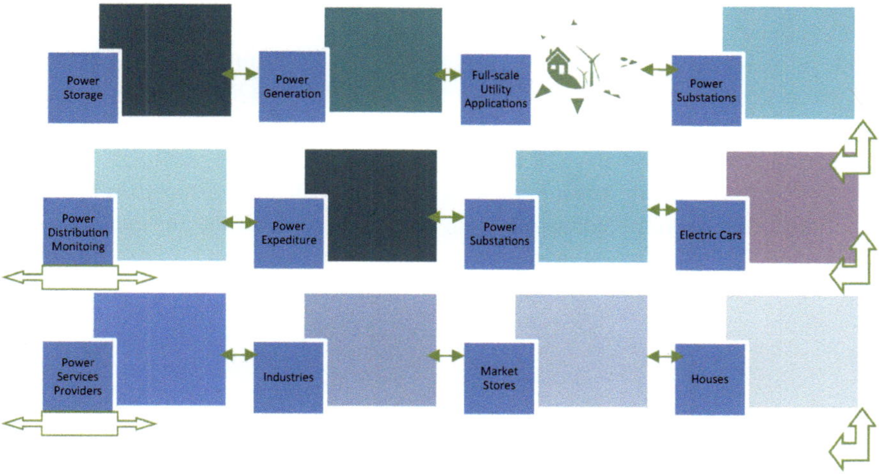

Fig. 5.6 An invented portrait of the smart grid

(SIG) is expected to self-control the whole power ecosystem status on both energy ends, that is users and providers alike (see Fig. 4.6). Endorsing the above authors reasoning, we further note that a customer is able to detect the quantity of power from his (her) residence, at the same time when the provider can control the capacity of energy used for clients (e.g., smart metering). Moreover, the SG provides with safety elements that can defense the network against faults or leakages. In a similar vein, the digital (computer-monitored) nature of SG against the electro-technical one of the Conventional Grid (CG) tends to rather differentiate the interaction (interface, communication) between the two grids: the first is expected to be bidirectional and the second one-way mode. In addition, the latter seems to reflect hierarchical modelling, while the first tends to denote a fully networked-based interrelated outflow-to-inflow (i.e., open innovation) energy flow infrastructure, providing end users with multiple power selection. In other words, summarizing the SG's inclusive features as proposed by the preceding authors, it seems to depict an "intelligent" power grid that can be comprised of smart metering, distributed (diffused) energy production, integrated renewable power resources, bidirectional interface, self-monitoring, data security, coal release restriction, energy demand and expenditure reaction and automated diffusion, among others. In particular, the dispersed energy production seems to "mirror" an integral part of the SGI, transferred as power generation capacity from undersized power resources, at time when the power plant power generation appears to relate to larger scale environmental impact on energy production and diffusion. Further, the distributed power generation tends to provide with market-oriented demand responses, ensure end-to-end secure power generation and diffusion, cost-effectiveness and protection from environmental disasters. Also, the renewable power integration embedded can advance the capacity (agency) of the grid to meet the energy demands in an appropriate response and ecologically-viable way. However, this environmentally-induced agency tends to be challenged by certain complexities (e.g., storage issues, how best to keep the solar power available at night or how efficiently embed both solar and airstream power resources to the Grid). Based on the aforementioned features, we outline that the prior authors conclude that the SG offers considerable merit at a larger scale than the CG. For instance, it is trustworthy, secure and can offer clean and renewable (environmentally-friendly) energy generation, transmission and storage for power consumers and providers as service clients and stakeholders, alike. Following their rationale, rare research evidence on Metaverse energy-promoted instructional immersive innovations and challenges for SG, Smart Factory (SF) and Metaverse-led sustainable energy L&D (professional practice) applicable across education and business is outlined next.

5.2 Metaverse Energy Battery-Learning Gaming and Development

Franco et al. [8] elaborate on their two learning games co-designed (i.e., open innovation) to facilitate postgraduate and citizen remote, interactive, immersive and engaging learning, development and engagement with battery energy and manufacturing education in France. The first learning game involved a mixed reality tool. The game users were assigned to be the drivers of a battery-led VR automobile designed to be refuelled, with their peers being in charge of managing the power generation, distribution, consumption and storage on additive manufacturing (i.e., 3D printing) appliances. The second learning game reflected a DT co-design solution. This tool depicted a DT lithium battery production line, which could be exercised either on Web or by employing a VR system. The authors claim that COVID19 pandemic has accelerated, among others, further attention on how best to communicate, work and socialize remotely using the emerging technologies. Along this line of enquiry, the exploration of Metaverse appears to be rather promising one. In other words, how we could interact and collaborate with others on evolving multimedia space(s) and devices (VR, AR, MR, XR) where virtual and physical objects, individuals, groups and organizations co-survive.

The authors reflect on their prior awarded research exploring previous VR learning game instruction on power-chemical battery domain and different battery technologies implemented on automotive industry [9]. In addition, they focus on another learning game of them, the "Smart Grid Mixed Reality" (p. 8). In this gameplay, the users assume the role of a power-led automotive driver in a VR setting, interacting with others (i.e., human co-players) on a 3D printed diverse power generator environment (i.e., airstream, sun, aqua). The gamers are guided to interact with physical objects in this immersive space in order to generate, produce and sustain virtual energy from wind, sun and (or) water resources, garnered, used and stockpiled for automotive power recharging. In this in-game context, the user(s) is(are) assigned to gather as many virtual offerings available, escape from corsairs' heavy vehicles and ensure that automotives are completely charged with energy. The driver of the automotive and three players that control the generation and storage of the various virtual power resources interact concurrently to achieve the gameplay assigned tasks. The authors reflect upon their effort to showcase that VR/MR learning games tend to demonstrate meaningful, immersive and engaging learning potential for energy generation, distribution and storage in battery science across different scholars and the public.

In their current paper, the authors elaborate further on their research upgrading their prior Smart Grid Mixed Reality (SGMR) learning game to that of the "Smart Grid Mixed Reality 2.0" (SGMR 2.0) (p. 11) one to engage 16 players in tandem. They first describe the game mechanics and user interface of this newer Mixed Reality (MR) learning game. Next, they report on their study exploring the range of learning motivation and cooperation attributes exhibited by the postgraduate students playing this game. Further, they bring in their newer VR Web-based learning game, called

"SIMUBAT 4.0". The latter reflects a DT of lithium battery manufacturing in their lab, as a seamless way to experiment with energy production, diffusion and storage on a collaborative Metaverse ecosystem.

The authors stress that exploring and exploiting the usage of renewable resources drives the successful conversion to sustainable society, with energy storage emerging technologies to be optimally used to adjust the process from renewable resource power generation production flow to expenditure as component of SG. Their newer learning game "SGMR 2.0" targets at showcasing the users' efficient smart energy generation, distribution, utilization and storage. One player is allocated in the VR in-game setting and 15 participants interact in the real space. The events occurring in the real environment tend to impact on the VR space and the gamer engaged in it. Within this setting, the player must ensure to drive the automotive by running away from corsairs' vans, not to run over stumbling blocks in the virtual environment and not to leave the automotive battery discharged. The driver needs to garner many awards when moving around the virtual city and its surrounding parts. These offerings seem to be present at a random order on the streets and their gathering provides the automotive driver with marks. When compared to their prior VR learning games and previous SGMR game, the authors illustrate that there seems to be no restriction in the amount of awards that can be garnered in-game.

Following an arranged number of offerings, the in-game levels are by design advancing. This means that more pirate vans come into sight and the obstacles that prevent road circulation will become more apparent. The automotive can be re-fueled by the gamer when parked for a short time in one of the four recharging stations in the virtual town. These stations are charged with the power stockpiled by the renewable (i.e., solar, wind, water) batteries and consumed by different units (e.g., home). The balanced equation between energy generation (from sun, airstream or hydro) and energy consumption (e.g., house) tends to depend upon the quality of interconnection between the batteries (different types of batteries) and the SG at a time at hand. When prompted to start the game, the VR user needs to select the category of battery and electric properties used in the automotive. Similarly to prior SGMR learning game, the driver can fuel his (her) car with one of three lithium-led batteries. The autonomous function of the automotive battery depends on the properties of the equipment becoming apparent through mathematical modelling in Python scripts. This type of modelling can be edited, therefore, the game can be expanded to any category of battery materials and technology, with the possibility of additional active extensions being made. When the automotive driver selects his(her) battery solution and related properties, he(she) begins driving the car around the virtual town and surrounding areas. The latter are dispersed along four interrelated sections that involve diverse types of housing, business, manufacturing and power production units. These residential houses, business constructions, plants and energy generators rest on airstream (wind turbines), solar power (solar furnace and cell panels) and hydro dynamic power (fluid barrier). A tsunami energy maker, which is completely managed by the learning game after a "real-world" giant sea wave life-cycle, is accessible at the aforementioned forth interrelated section. The above residential buildings, businesses and manufacturing entities use power, with their energy levels monitored through 3 3D-printed appliances that represent a home, a workplace construction and an industrial unit, accordingly. The power expenditure levels in these machines are monitored through controllers configured via Python scripts by three users (gamers) in the real environment. The aforementioned appliances are linked via Bluetooth to a PC. The participants have to launch power expenditure at diverse levels across the four sections indicated above, pending on their selected strength and the amount of buildings accessible in each of the above four sections. This element seems to showcase by and large the power expenditure range for each of them. The gamers performing on these buildings earn more awards if they use more power, through an automated in-game analytics feature. The SGMR 2.0 learning game was preliminary implemented as a team-based lab classroom activity. The aim was to engage the academic tutor and the postgraduate students into multi-level and multi-reality immersive active in-gaming experimentation to improve their L&D on Metaverse-linked energy (battery) generation, expenditure, dissemination and storage successfully. The in-game design included 5 users interacting with 5 3D-printed devices, with one participant acting on a 3D-printed solar furnace and another peer interacting with a 3D-printed fluid tube. These energy generation appliances are linked via wireless transmission to the PC replicating the actual world. The energy generation range in each of the aforementioned four sections rests on the category of appliance employed, the mode of usage (e.g., the strength of the beam utilized to light up the solar furnace, the similar one of the airstream on the corresponding coil apparatus, etc.) and their comparative amount in each of the four sections. When attendees act on them, these appliances launch power production in the VR setting. The power generation range can be configured through Python scripts which can also address complementary engineering components of the airstream devices. A climate simulator (representing real-world cycle between rainwater, solar activity and airstream) offers active indicators through control panel on which category of power device to employ at each spot of the game. The users performing on the appliances need to do their best to keep to these indicators in order to garner as many awards as they can. The in-game design automatically counts these offerings as a way of operation between the categories of the power appliances employed and those required by the climate control panel. Four gamers interact with actual buttons ("switches" p. 9), which can take on "unlocked" or "locked" status. These buttons act at the line between Sects. 5.1 and 5.2, 5.2 and 5.3, 5.3 and 5.4, and 5.4 and 5.1. By setting a button between two sections in the "unlocked" status, power runs from the section with an extended number (quantity) to the one having less. This may be employed to reimburse for the shortage of power in a particular section. Appropriate power diffusion between sections (i.e., confirming that none of the sections is left without power) permits the "switch" (p. 10) users to garner in-game awards, automatically. The buttons act through wireless transmission on the VR setting replicated by the PC. At a final point, a user acts on an actual "Chaos" (p. 9) helm also linked through wireless transmission with the PC in which the VR condition is replicated. By rotating the helm, it will accidentally end to one of the various pointers causing virtual disasters: this entails the immediate complete dismissal of the inactive battery, the buttons ("switches" p. as above) which freeze for a number of seconds (mirroring challenges in the energy diffusion ecosystem), the airstream turbines, the solar furnace or the fluid barrier which can't generate energy for a number of seconds. The helm can also end at a corner where any disaster is formed. The appearance of these disasters drives the users to re-create the approaches used to keep the appropriate equilibrium between power generation, expenditure, diffusion and stockpile. The total number of garnered awards is estimated in-game, by the amount of offerings gathered by each individual participant performing on the various 3D-printed items.

The SGMR 2.0 game can be employed across two potential conditions: antagonistic one between individual agents or groups that are assigned with specific tasks (e.g., power production, diffusion, expenditure) or cooperative one targeting at confirming the energy sustainability of the grid. In this game adaptation, the awards garnered by the automotive driver through the offerings he(she) gathers do not corroborate to the total number of awards gained. Even so, if the car driver escapes his(her) duty (either due to corsair hostage or barrier collision or complete automotive battery run out), all the gamers lose. The game seems to bear no definite time interval. It can go on many hours, resting on the users' progress, but the assigned students played the game for approximately 120 min. The game permits keeping an evolving game activity for a later interval, therefore, the users are able to manage their approach and handle exhaustion as they prefer.

The authors argue that since they co-designed the SGMR 2.0 game, they employed it on a regular basis in classroom T&L for Master of Science (MSc) power adaptation and depository at a French university, science week events and particular demos to guests of their institution. In their study exploring the pedagogical effect of the above game, they examined the gameplay learning outreach, participants' motivation, cyber-fatigue, flow, immersion and social presence, learners' experience and shared cooperation. Following the first author's 8 tutorials lasting 90 min each on emerging technologies for power transformation and maintenance, a team of 8 learners enrolled at the MSc Power Maintenance and Adaptation curriculum were prompted to gain the highest number of awards by both running the MR power grid ecosystem and by garnering as many offerings they could in the VR setting. The learners were required to fill in a number of questionnaires pre- and post-gaming. The first pre-gameplay measurement tool included 19 items: 3 of demographic information (age range, legal gender, national origin), 9 adapted statements about ease with emerging technologies (VR, AR, MR) and 7 questions about acquaintance with power generation and renewable energy resources. The additional assessment tools used explored learner motivation, simulation fatigue and subject knowledge. The self-reported questionnaires used after gaming explored the learners' usability experience, immersion aspects and presence. After completing the pre-assessment tools, the participants were introduced to the game by the first author. Next, they practiced the game and once finished with gaming, they filled in the after-gameplay assessments. The authors recorded the users' interaction and dialogues while gaming. The learners appeared to be VR-technologies savvy. However, there seemed to be no significant difference in awareness of energy conversion and storage pre- and post-gameplay. Likewise, pre- and post-student motivation did not indicate any significant difference. The game was reported not to create cyber-fatigue. In addition, it seemed to convey immersion (focus, feeling, gamification) and a sense of presence (impression of real-world, performing action, user interface iteration and self-evaluation), followed by user interface attractiveness and edutainment. Further, the game seemed to trigger engagement in collaborative activities between and within the first author and the students, depending on their in-game roles. The authors conclude that although the SGMR 2.0 game appeared not to rather advance the users' knowledge and practice of energy conversion and storage subject, it seemed to bring meaningful immersion to them and improved sense of presence and user interface experience. Noteworthy tends also to be the collaborative nature of the game, dispersed along the in-game sub-levels, with improved information exchange in order to achieve the highest potential game score.

The authors further elaborate on their newer co-designed SIMUBAT 4.0 (p. 10) game to act on a lithium- or sodium-ion battery (p. as above) production process simulated in a VR setting. Such replication is expected to be a DT of the home battery manufacturing at the authors' Solid Chemistry university courses. In the gameplay, the users are guided to follow the expected steps to complete the game in a Global Positioning System (GPS) similar manner. At the start of the game, the user needs to visit the designated dashboard to go through the lithium ion battery cell properties and power indicators which are configurable. Next, the user is assigned to manufacture a black lead wire (positive or negative) across sub-level gaming representation steps with the usage of a control panel, followed by the fabrication of battery cells with lower power consumption. The authors co-designed an improved version of the game available on the Web to familiarize the public with the battery cells manufacturing in a flexible, remote and self-directed mode. The authors reflect upon their co-designed immersive solutions where interacting with interrelated devices through a laptop with the VR setting manipulated by ongoing input from appliances managed in the real environment, can drive towards multimodal T&L and collaborative immersive spaces. They stress the option of diverse learners, academics, researchers and practitioners working remotely across different areas on power conversion and storage L&D through immersive technologies. For instance, a group of individuals in one region can perform on the 3D-printed airstream turbines, with another one acting on the power circulation switches and another team managing the automotive driving in the VR setting. They claim that apart from its pedagogical meaningfulness, the SGMR 2.0 game seems to be an active multimodal simulation system that spans across the fabrication of battery cells to the automotive and the energy grid conversion. In other words, it is expected to be applied to explore complicated and configurable connections between battery solutions, car driving and exchange between power production, circulation, expenditure and storage. In this way, it can be used as research tool, with scholars experimenting with diverse battery cell fueled modelling, airstream to electricity generation or battery-charging models effectively. In addition, the authors illustrate that the IoT design of their game facilitates open, collaborative and transdisciplinary research engagement by drawing together electrical engineering and materials experts to cooperate remotely through a MR setting while being in different sites. Their interactions with physical objects and devices in the real environment with the VR setting promoting the true interconnection between real and virtual world, with them playing the game(s) supporting scientific research development further. The SIMUBAT 4.0 (p. as above, Fig. 5.8) game might be expanded to additional power-led subject areas to promote transferable skills (e.g., how best to address chemicals safely) in a low risk simulated L&D context that breeds VR flexibility and configurable instructional design (i.e., materials, objects, devices, processes may be inserted following adapted needs, Figs. 5.7 and 5.9 next).

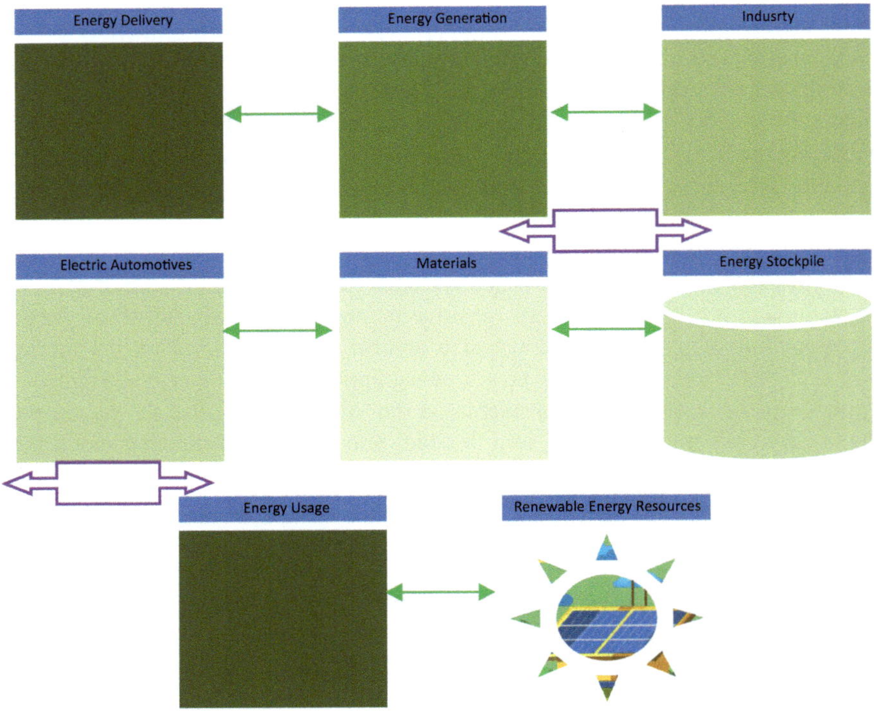

Fig. 5.7 The smart grid mixed reality 2.0 learning tool operationalization

Fig. 5.8 The SIMUBAT Metaverse learning game concept

Fig. 5.9 The Metaverse collaboration space notion

5.3 Smart Factory Learning and Development

Alpala et al. [10] argue that the advanced virtual reality tools eminent in recent years have shifted the ways we communicate, learn and develop (either in higher education or industry) during an immersive "real-world sense" experience, without the risk of serious error(s) consequences. In their current work they elaborate on their attempt to optimize the interaction experiences in 3D virtual environments that blend VR and Metaverse platform in the context of a smart plant. A core insight of their rationale seems to be that the traits of the real-life setting can be represented in a virtual space through DT and avatars (p. 4). Along this line, they aim to report an experimental model that assumes the introductory stage in the procedure of optimizing cooperation in virtual settings through VR-fused Metaverse ecosystems. The latter consist of operational elements, object-adopted designs and enhanced multiple-user applications. The authors describe their study of the first implementation case study of VR in Metaverse ecosystem replicating a smart factory (p. 3). Their Metaverse-based smart plant system was assessed with end users through diverse user experience indicators that showcased the promising evidence of its possible L&D (training, educational) and business application. This tends to lend support to the potential power of merged immersive technologies for industrial integration in favour of emerging, shared work-based learning, training, teamwork and communication interactions. In order to design their system, the authors employed the Unreal Engine 4 (i.e., game development) approach shifted to VR and Metaverse-promoted design. In their final section, they report on their user experience assessment that indicated favourable evaluation of the Metaverse-based smart factory system. The researchers argue that the Metaverse seems to progressively advance its status as an extended business workstation, globally. An interesting predictable effect is that enterprises will develop digital replicas of their procedures (aka DT), which can be assessed in the Metaverse space before being physically applied. In that respect, the layout can accept developments pre-design phase, being cost-effective and time saver. One of the critical features that constitutes the design of the Metaverse reasonable is the high immersion indicated by VR tools, which permits clients to interact with diverse devices and digital objects like in the physical environment. In that sense, the integration of Metaverse into everyday social and professional scenery will be connected to the configuration of the VR solutions. Lastly, the related so far work reveals the promising future of the VR-fused Metaverse, an "alternative future" that in the next years will be gradually positively and extensively accepted.

Metaverse Smart Factory (Manufacturing Plant)

The authors implemented tailored iterative design principles based on immersive virtual environments characterized by advanced visualization and real-world exercise conditions that improved the scalability of design and assessment session(s) for the L&D (or training) of workforce in actual time, employing the 3D-Metaverse VR platform. Their embedded system took after the name of "Digital Factory Metaverse" (p. 11). Their factory architecture includes the production lines, the raw resources, the final products and supplementary operational functions and utilities. In addition, their system was configured to be performed in remote, multiple-user version where teams of customers can visit the plant and experience the production process resembling the real-life one. The research participants consisted of 9 female and 11 male engineering students ($N = 20$), being enlisted through support from a higher education institution. In order to perform the user experience assessment, the respondents formed 3 teams based on their prior practice of VR applications. Before group formation, each participant was provided with preliminary training for the usage of the VR system. The assigned classification of the aforementioned groups was the following, as next. The first team included 8 participants with no previous experience of VR tools. The second group involved 6 attendees with prior practice of gaming, but not VR. The third and last team consisted of 6 respondents that exhibited some prior practice with VR systems. Each team was allowed to experiment with the VR system through different sessions moderated by a tutor for approximately 2 h. The authors run their system with the MetaQuest 2 and Oculus Rift VR HMDs, controllers and PCs with specifications.

Outcomes

The authors first deployed the application of the VR solution in a production plant with a smart factory method (p. 13), for which a confirmation of the approach and the design strategy of the creation of VR tools was performed. Further, the system was certified as a Metaverse-based design in remote multiple-user version with a team of clients in group partnership, the application of avatars and user experience assessment through indicators. In addition, the authors attempted to achieve advanced Metaverse-facilitated teamwork immersion by empowering the elements of the sense of being there (i.e., that of presence) and feeling of real-world replication (i.e., realistic depiction) in their scalable system. As regards the former feature, engaged participation was fostered by an array of visual signals, audio, text and mock-ups that could assist them to handle the data in actual time. In relation to the latter, the engine used embedded screenshots, objects and features that supported navigating around the platform in a "real-life" and interactive scenery effectively.

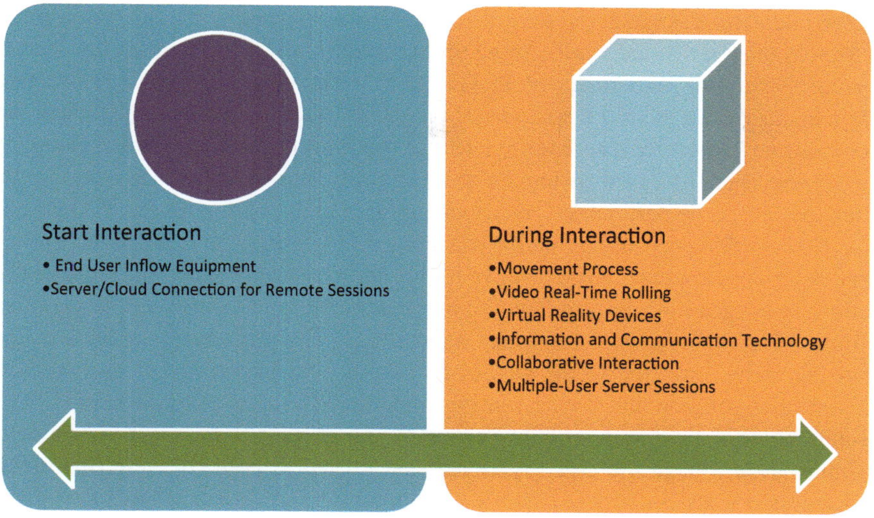

Fig. 5.10 An outlined reflection of the VR Metaverse smart factory design system

The Smart Plant

The authors developed their solution of the "smart factory" (p. 14) using the VR system approach and according to their suggested methods design. In that respect, they designed a fully-fledged virtual plant with appropriate services, apparatus, production lines, storehouses and other elements that framed the overall factory business. The technologies used varied from AI and DT to cloud computing, data analytics, collaborative workspaces and assembly lines, respectively. A preliminary user experience system assessment was performed to reveal potential issues for development. A VR-Metaverse reflection of their designed system is sketched in Fig. 5.10.

The authors identified four types of user experience assessment indicators of the VR solution, as next. First, the user practice. Second, the monitoring of the devices. Third, the multiple-user interface. Fourth, the immersion experience and the time allocated to complete the interaction.

The remote multiple-user VR application was effectively assessed with teams of end users who practiced the system under matching circumstances. The core statements were assessed for user practice spanned across L&D, system usefulness and perceived analogy of the VR solution with other soapboxes, visualization tools, traversing and interface practice. After the introductory user assessment sessions, 7 developments were suggested and adopted to be integrated into the system. First, to support attendees feel a higher level of the sense of being there, to perceive the VR context as that of real-world environment and experience a greater sense of interaction within the virtual setting. Second, to facilitate the participants have a higher sense of monitoring their fluency with the avatar-agents. Third, to foster respondents experience an elevated level of data usage. Fourth, to help the attendees receive better interaction assistance during the system's experience. Fifth, to increase the sense of being in the real world during interaction. Finally, to strengthen the "hands on" L&D (training) experience on Metaverse-based ecosystem. 20 male and female participants took part in the assessment procedure. 88% of the male attendees did demonstrate favourable responses in terms of user interface practice and usability of the system, with 70% of their female peers reporting positive reactions. Among the attendees, the male participants allocated approximately 2.8 h with the VR system, with their female peers assigning 2.4 h, accordingly. Despite the fact that the overall user experience appeared to be positive for the attendees, yet the authors indicate the following suggestions for development in the newer versions of the VR solution, as follows. The Meta Quest 2 headgears appeared to offer participants more comfort in using them compared to the Oculus Rift ones, possibly attributed to increased lenses' configuration and higher sensing integration for movement detection. The avatar-agents of the system would be customized according to the users' recommendations. Even though the user experience and related assessment were performed in lab sessions, the authors emphasize the future in-house usage of the VR system. The participants with no prior practice with VR tools seemed to experience some challenges with controlling the interface interaction during navigation, compared to their peers with some previous experience with VR or gaming. Having said that, the authors expect that end users will be prepared and (or) ready to use the system based on greater familiarity with the solution. Their questionnaire used for user experience assessment was designed to explore multiple aspects of usability of the system and participants' attributes during interaction (i.e., reflecting user practice, VR readiness, multiple-usage and immersion/flow). The attendees who had prior practice with VR tools seemed to need lowered readiness and trial time. 95% of male participants and 89% of female ones appeared to be more adaptable to prepare for and screen the VR equipment. The attendees who used the VR solution with no previous experience, their initial familiarization time was greater, while their subsequent use of the VR apparatus was more handy than before. 93% of the male and 90% of the female participants reported a satisfactory practice when employing the multiple-user delivery both at the interaction stage and team cooperation. In particular, the multiple-user configuration seemed to be very helpful for teamwork on Metaverse platform, as all attendees confirmed realistic experience throughout. In terms of immersion assessment indicators and outcomes, it seems that the participants with no previous experience with VR systems and gaming, reported minor tiredness, with their VR experienced peers revealing no cognitive load challenges. The authors stress that any slight tiredness potentially associated with VR system usage might well vanish after recurrent use.

In outlining their arguments, the researchers emphasize the fact that VR and Metaverse-supported technologies are gradually entering our education, L&D (training, professional practice) and business environments, permitting clients, employees or learners from any region in the world to join the immersive workspaces. In addition, they add that the practical challenges related to the current VR Metaverse-supported solution might be well implemented within any industrial sector without higher expenses and (or) increased challenges met due to limited L&D. Their obtained user experience outcomes seem to showcase the dynamics of VR Metaverse-facilitated systems for the first time across multi-level environments. Especially for L&D purposes, the VR Metaverse ecosystem can support employee training before performing it in the real workplace. In that respect, the development of Metaverse platform seems to be a rather lowered expenses tool for enterprises and (or) learning organizations. Further "saving" the environmental and financial impact of travelling and possession of equipment and resources, since trainees can execute training exercises as frequent as they need, before getting into actual attempts. Finally, the authors recommend further developments over newer related research and work to strengthen the adoption of Metaverse-based systems in diverse production L&D business settings.

5.4 Campusverse Engagement Challenges with Teaching and Learning-Learning and Development

Onu et al. [11] elaborate on their qualitative exploration of benefits and challenges associated with the XR Metaverse (aka Campusverse) immersive and experiential teaching and learning infrastructure. Their work seems to emphasize the promising dynamics of immersive and personalized education remote provision through the Metaverse instructional ecosystem. The researchers performed a number of interviews to garner learners, supervisors and programme officers perceptions and insights on Metaverse-led teaching and learning aspects. Their evidence suggests that the effective endorsement of the large-scale Metaverse learning platform brings forward the attention and (or) the need to address various Campusverse-interrelated technical, cultural and ethical challenging features (e.g., engaged connection, reliability, standardized certification, enterprise operational preparedness and context-specific knowledge and skills) to harness the Metaverse's ecosystem expected rewards. The Metaverse higher education institutions that resemble business operations, need to ensure that their workforce, their learners and other engaged stakeholders can perform their academic, teaching, research, administration and innovation or knowledge transfer responsibilities in a technologically advanced, open, networked, distributed but secure, reliable and data protective mode. This Metaverse for higher education infrastructure is expected to embrace XR-led classrooms, R&D simulations and fabrication laboratories, among others. Within and across this emerging platform, learners and tutors are expected to share their interests and expertise on global collaborative research, resourcing and interaction conditions. The technology related to Metaverse seems not to have been widely integrated across the world due to diverse challenges [e.g., issues of technology adoption and (or) access due to cost effectiveness or other constraints, etc.]. However, the relative debate on Metaverse services and education-powered seamless spaces that merge the physical with the virtual environment in real-time (e.g., AI/AR/VR/MR/XR, HMD, tracking and holographic controllers, mobile devices, etc.), in particular, is evolving in range and frequency as solutions are becoming more inclusive and of lowered price. Following this expected outflow to inflow networked integration (i.e., open innovation) of physical and digital objects, real-life analytics and human agents' expertise and exercise, the multimodal irreproachable Metaverse education is expected to launch and sustain the co-creation (i.e., open innovation), sharing (transfer) and maintenance of evolving global knowledge and skills across space and time.

The authors elaborate on their interview-based exploration of a Metaverse-led academic programme engagement across diverse software developers, learners, supervisors, managers and curriculum officers. 28 experts in Metaverse-based education (e.g., AI/VR/AR/MR/3D games, holographic solutions, etc.) around the world shared their experience of implementing Metaverse-led teaching and learning compared to conventional instructional modes and their views on future usage of Metaverse-linked teaching and learning. Additional data from relative reports were also integrated into the research design performed. The aim was to capture further standpoints on the subject of Metaverse-based education that covers a new evolutionary notion. The interview questions were based on literature review performed, showcasing limited yet growing research interest on Metaverse-triggered education. The researchers emphasize the contribution of the participants' diverse professional background organizational roles (e.g., senior technology advisors, sales and business development managers, professors of AI, graduate and postgraduate students, among others, etc.). 21 of the attendees seemed to exhibit a favourable attitude and exercise with Metaverse-associated instruction. Their other 7 peers tended either not to be rather positive towards Metaverse-led education or not sure of its capacity and range of support as far as higher education or industry L&D are concerned. The majority of the attendees demonstrated considerable concern and eagerness to embed Metaverse into teaching and learning. They elaborated on how its extraordinary immersive and exchange character can support greater learner engagement and facilitate advanced learning outcomes. In addition, the participants stressed the dynamics of personalized and adaptable learning agency of Metaverse. They illustrated its capacity to provide students with personalized and self-directed experiential learning further to its potential to integrate diverse learning types. The associated Metaverse-reported challenges raised matters about security of the ecosystem, accessibility concerns and possible interruptions within the Metaverse platform. In addition, the respondents raised the following issues of the Campusverse future adoption, as next. First, the operating costs related to establishing and sustaining such an engaged open-led digital innovation network and intersectionality. Second, the regularity and accreditation process relevant to Metaverse-based technology implementation. Third, the standardization of data privacy and security required for Metaverse education. Fourth, the flexibility in overcoming potential obstacles to business preparedness for implementing the Metaverse-linked teaching and learning. Fifth, the transdiciplinary professional expertise to co-design (i.e., open innovation), use and sustain the Metaverse-associated solutions in higher education and (or) industry. Overall, the authors conclude that handling the above pitfalls seems to be vital to safeguard the future efficient and inclusive realization of Metaverse outflow to inflow

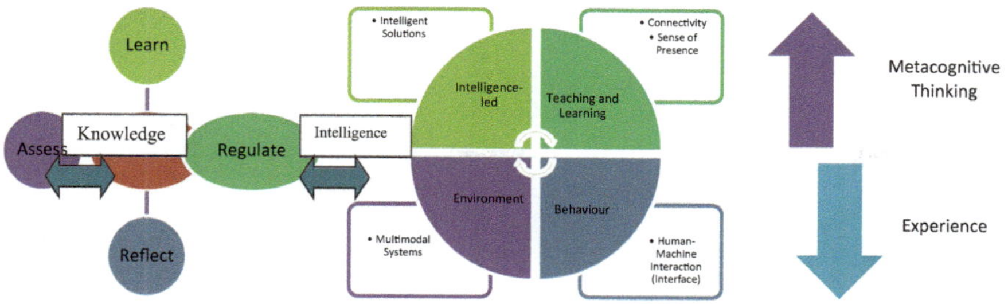

Fig. 5.11 An integrated open innovation-led Campuseverse and business learning

(i.e., open innovation) academic and business teaching and learning, depicted in a visualized integrated outline in Fig. 5.11.

Chamola et al. [12], Yaganeh et al. [13] and Stripan et al. [14] elaborate on benefits and challenges that future education (including adult or further education) might bear as transferred on Metaverse. Reflecting on the above authors reasoning, we argue that they seem to rest upon Metaverse's potential impact on curriculum learning experience (e.g., cross-over disciplines: manufacturing, business, energy supply chain management built on lowered waste of physical resources, etc.), spanning across the Metaverse "Smart Education Complex", as a multimodal, systems dynamic modelling (i.e., open innovation) approach. This "mirroring", intelligence-based operationalization of Metaverse Smart Education Ecosystem is further triggered by its immersion, expected to complement the pitfalls of the physical learning environment, even able to overcome the restrictions of the physical education space in certain aspects. The preceding authors outline that the majority of current Metaverse-related work tends to focus on a particular branch (system) of technology. Despite the fact that this development tends to validate the fundamental overall Metaverse technology integration paradigm, it seems that it hasn't supported the inclusive, industry-and-education intelligence across diverse sub-sector Metaverse infrastructure, successfully. In summary, we reflect on the authors concluding insights by indicating the following points, as next. First, the request to introduce the notion of Metaverse in diverse teaching and learning (e.g., academic and business programmes) tends to be rather more evident or greater in certain disciplines (e.g., life sciences, healthcare, manufacturing, urban development, etc.). Second, the corresponding demand for engineering and geoengineering sciences tends to be rather less evident or robust. Third, the introduction of Metaverse-based concept and exercise across interdisciplinary curricula in the next years, is expected to evolve more co-designed (i.e., open innovation), inclusive and fusion-Metaverse curricula that could foster a more "intelligent" (smart)-led optimized tertiary and business education. Moving forward the authors' representation of a re-generative Metaverse teaching and learning ecosystem, we sketch the future open innovation (i.e., external to internal flow of data, processes or interaction) Metaverse-led higher education and business infrastructure in the following flowchart (integrated Fig. 5.12).

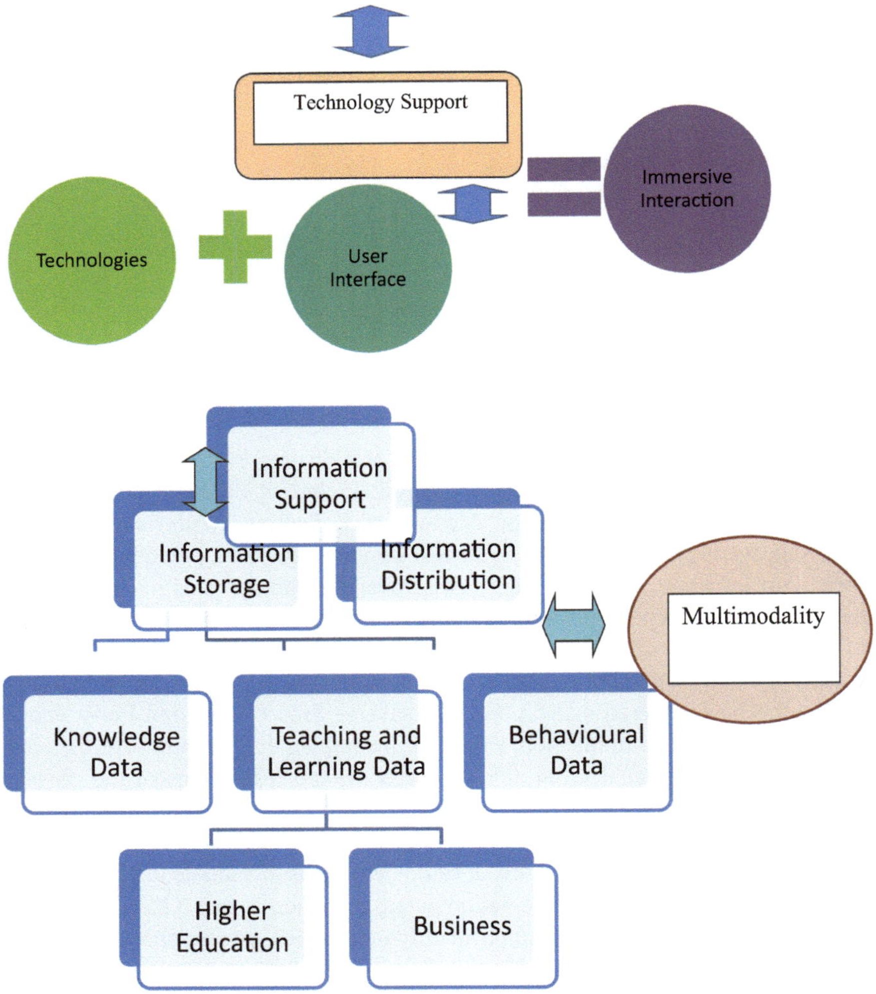

Fig. 5.12 An outline of the Metaverse higher education and business systems dynamic modelling

Our aforementioned outline of the expected future Metaverse-led education space denotes the global business and (or) education-led teaching and learning developments in moving towards sustainable immersive intelligence, inclusion, exchange and open innovation. In that respect, this Metaverse learning ecosystem integrates not only smart intelligence tools to support users' curriculum, service learning and study management optimization, including that of smart energy education and (or) L&D, but also integration of in-and-out-classroom learning and accreditation system(s).

5.5 Conclusion

Apart from the presented authors above, growing literature (e.g., [15, 16]) introduces and explores the notion and significance of Metaverse teaching and learning in adult and professional practice instruction. Following-up on their summarized reflections, we consider the following directions, as next. First, the smart intelligence of Metaverse mature education research and prospective exercise is expected to rise and become the key driver for next re-generation, next frontiers higher education in the years to come. Second, tertiary education and business L&D need to plan how to co-design (i.e., open innovate) the structure or framework of the envisioned Metaverse-led L&D paradigm. Third, the assessment and fine-tuning of Metaverse-based instruction must reflect upon the open innovation (i.e., external to internal flow of data, information, processes, services, accreditation, regulation, etc.) engaged approach. Fourth, the introduction of new Metaverse-based features into transdisciplinary, energy-led greener domains in higher education and (or) enterprise are expected to improve the quality of education offer by infusing transformative (i.e., re-generative) and real-life learning and professional sustainable practice support [17].

Our suggested assumption is that the evolution of the Campusverse open innovation (i.e., external to internal flow of R&D networks) ecosystem is expected to leverage XR and quantum solutions to re-generate (i.e., re-invent, re-innovate) and sustain higher education and enterprise upskilling (or reskilling) and growth, efficiently. It is envisaged that through this emerging disruptive transformation, the tertiary education and business institutions are expected to be able to accomplish personalized teaching, learning, assessment and mentoring, greater sustainable (greener) operations, learner equity, diversity and inclusion (EDI), higher resource-led cost efficiency and more resource-driven eco-friendly summative functionalities.

Endorsing this evolving Campusverse prototype, a deep but balanced context-specific and not currently "immature" education-based investigation of, reflection upon and debate on possible risks and (or) challenges should be launched. This exploration should align with, for example, learning and operational performance, workforce restructuring and ecosystem's privacy, safety, interoperability and maintenance based on eco-innovation principles (or properties). This envisioned open innovation-driven sustainable Campusverse reflection might be equally explored within both higher education and enterprise infrastructure, as well. The conceptual flowchart of our reflective operationalization of the predicted Campusverse-powered, open innovation-led ecosystem is illustrated in Fig. 5.13, as next.

Fig. 5.13 The Campusverse open innovation ecosystem

References

1. Tan Y, Xu W, Li S, Chen K (2022) Augmented and virtual reality (AR/VR) for education and training in the AEC industry: a systematic review of research and applications. Buildings 12:1529. https://doi.org/10.3390/buildings12101529
2. Yaqoob I, Salah K, Jayaraman R, Omar M (2023) Metaverse applications in smart cities: enabling technologies, opportunities, challenges, and future directions. Internet Things 23:100884. https://doi.org/10.1016/j.iot.2023.100884
3. Sin ZPT, Jia Y, Wu ACH, Zhao IDZ, Li RC, Ng PHF, Huang X, Baciu G, Cao J, Li Q (2023) Towards an Edu-Metaverse of knowledge: immersive exploration of university courses. IEEE Trans Learn Technol 16(6). https://doi.org/10.1109/TLT.2023.3290814
4. Hare R, Ying Tang Y (2023) Hierarchical deep reinforcement learning with experience sharing for Metaverse in education. IEEE Trans Syst Man Cybern 53(4):2047–2055. https://doi.org/10.1109/TSMC.2022.3227919
5. Janev V, Berbakov L, Tomašević N, Sotoca JMB, Lujan S (2025) Validating the smart grid architecture model for sustainable energy community implementation: challenges, solutions, and lessons learned. Energies 18:641. https://doi.org/10.3390/en18030641

6. Yadav S, Anand R (2025) Analysis of advancing paradigms of smart grid innovations, applications, challenges, future trends and strategic implementations. Discov Appl Sci 7:1380. https://doi.org/10.1007/s42452-025-07905-2
7. Kabeyi MJB, Olanrewaju OA (2023) Smart grid technologies and application in the sustainable energy transition: a review. Int J Sustain Energy 42(1):685–758. https://doi.org/10.1080/14786451.2023.2222298
8. Franco AA, Loup-Escande E, Loiseaux G, Chotard JN, Zapata-Dominguez D, Ciger J, Leclere A, Denisart L, Lelong R (2023) From battery manufacturing to smart grids: towards a Metaverse for the energy sciences. Batteries Supercaps 6(e202200369):1–15. https://doi.org/10.1002/batt.202200369
9. Franco AA, Chotard JN, Loup-Escande E, Yin Y, Zhao R, Rucci A, Ngandjong AC, Herbulot S, Beye B, Ciger J, Lelong R (2020) Entering the augmented era: Immersive and interactive virtual reality for battery education and research. Batteries Supercaps 3:1147–1164. https://doi.org/10.1002/batt.202000120
10. Alpala LO, Quiroga-Parra DJ, Torres JC, Peluffo-Ordóñez DH (2022) Smart factory using virtual reality and online multi-user: towards a Metaverse for experimental frameworks. Appl Sci 12:6258. https://doi.org/10.3390/app12126258
11. Onu P, Pradhan A, Mbohwa C (2024) Potential to use metaverse for future teaching and learning. Educ Inf Technol 29:8893–8924. https://doi.org/10.1007/s10639-023-12167-9
12. Chamola V, Hassija V, Singh A, Mittal U, Pareek R, Mangal P, Devika D, Mahmud M, Brown D (2023) Metaverse for education: developments, challenges and future direction. Preprints: 2023081872. https://doi.org/10.20944/preprints202308.1872.v1
13. Yeganeh LN, Fenty NS, Chen Y, Simpson A, Hatami M (2025) The future of education: a multi-layered metaverse classroom model for immersive and inclusive learning. Future Internet 17:63. https://doi.org/10.3390/fi17020063
14. Sripan T, Jeerapattanatorn P (2025) Metaverse-based learning: a comprehensive review of current trends, challenges and future implications. Contemp Educ Technol 17(3):ep584. https://doi.org/10.30935/cedtech/16434
15. Zi L, Cong X (2024) Metaverse solutions for educational evaluation. Electronics 13:1017. https://doi.org/10.3390/electronics13061017
16. Kurniawan Y, Susandyoga AE, Kamal IA, Sismandrajaya R, Hiererra SE, Bhutkar G (2023) Analysis of readiness to use the metaverse platform in learning activities. Emerging Sci J 7(6):2133–2155. https://doi.org/10.28991/ESJ-2023-07-06-016
17. Abdelmagid AS, Jabli NM, Al-Mohaya AY, Teleb AA (2025) Integrating interactive metaverse environments and generative artificial intelligence to promote the green digital economy and e-entrepreneurship in higher education. Sustainability 17:5594. https://doi.org/10.3390/su17125594

FSC
www.fsc.org
MIX
Papier | Fördert
gute Waldnutzung
FSC® C083411

Kolibri 360 GmbH
Ferdinand-Jühlke-Straße 7
99095 Erfurt, Deutschland
produktsicherheit@kolibri360.de